WASTEWATER TREATMENT FUNDAMENTALS II

SOLIDS HANDLING AND SUPPORT SYSTEMS

OPERATOR CERTIFICATION STUDY QUESTIONS

2021

Water Environment Federation
601 Wythe Street
Alexandria, VA 22314–1994 USA
http://www.wef.org

Association of Boards of Certification
2805 SW Snyder Blvd., Suite 535
Ankeny, IA 50023
http://www.abccert.org

ISBN: 978-1-57278-361-4

Water Environment Research, *WEF*, and *WEFTEC* are registered trademarks of the Water Environment Federation.

About WEF

The Water Environment Federation (WEF) is a not-for-profit technical and educational organization of 35,000 individual members and 75 affiliated Member Associations representing water quality professionals around the world. Since 1928, WEF and its members have protected public health and the environment. As a global water sector leader, our mission is to connect water professionals; enrich the expertise of water professionals; increase the awareness of the impact and value of water; and provide a platform for water sector innovation. To learn more, visit www.wef.org.

About ABC

The Association of Boards of Certification (ABC) was founded with the mission to advance the quality and integrity of environmental certification programs throughout the world. This charge has held strong through more than 45 years of providing knowledge and resources to nearly 100 certifying authorities representing more than 40 states, 10 Canadian provinces and territories, and several international and tribal programs. ABC believes in certification as a means of promoting public health and the environment while striving to give our members the necessary tools to ensure the knowledge and skills of their operators.

Prepared by

Sidney Innerebner, PhD, PE, CWP, PO, Indigo Water Group, *Contractor*

Dark Water Solutions, Ltd., *Contractor*

Chris Maher, *Coauthor*

Stacy J. Passaro, PE, Passaro Engineering, LLC

Kim Riddell

Kenneth Schnaars, PE, *Coauthor*

Bill Snyder, President of Snyder Technical Services, LLC

Roger Zieg, *Coauthor*

With assistance and review provided by

Paul Bizier

Bill Brower, PE

Jeanette Brown, PE, BCEE, FWEF

Raj Chavan

Frank DeOrio, PO (ABC)

Houssam El Jerdi

Stephanie Fevig

Richard Finger

Mike Gosselin, PO (ABC)

Georgine Grissop, PE, BCEE

Maria Dolores Guerra

Marialena Hatzigeorgiou, PE

Eric Holmquist

Ryan Hurst

Vishakha Kaushik, PE

Paul Krauth, PE

Rick Lallish

(Emy) Wenxin Liu, PhD, PE

Shaun Livermore, PO (ABC)

Jennifer Loudon

Karthik Machala

Indra N. Mitra, PhD, PE, BCEE

Andy O'Neil, PO (ABC)

John R. Reynolds, PO (ABC)

LeAnna Risso, PO (ABC)

Greg Seaman (ABC)

Ajay Shrivastav

José Velazquez, PE, BCEE

Christine Volkay-Hilditch, PE, BCEE

Steve Walker, CWP

Paul Wood

Paula Zeller

Emily Zidanic

Under the Direction of the **Technical Practice Committee**

Andrew R. Shaw, PhD, PE, Chair

J. Davis, Vice-Chair

Dan Medina, PhD, PE, Past Chair

H. Azam, PhD, PE

G. Baldwin, PE, BCEE

S. Basu, PhD, PE, BCEE, MBA

M. Beezhold

P. Block, PhD

C.-C. Chang, PhD, PE

R. Chavan, PhD, PE, PMP

M. DeVuono, PE, CPESC, LEED AP BD+C

N. Dons, PE

T. Dupuis, PE

T. Gellner, PE

C. Gish

G. Heath, PE

M. Hines

M. Johnson

N.J.R. Kraakman, Ir, CEng

J. Loudon

C. Maher

M. Mulcare

C. Muller, PhD, PE

T. Page-Bottorff, CSP, CIT

T. Rauch-Williams

V. Sundaram, PhD, PE

M. Tam, PE

Contents

Chapter 8   Aeration Systems 73

Chapter 9   Laboratory Procedures 85

Chapter 10   Chemical Storage, Handling, and Feeding 101

Association of Boards of Certification Formulas

Wastewater Treatment, Collection, Industrial Waste,
& Wastewater Laboratory Exams

---

*Pie wheel format for this equation is shown at the end of the formulas.

$$\text{Alkalinity, mg/L as CaCO}_3 = \frac{(\text{Titrant Volume, mL})(\text{Acid Normality})(50\ 000)}{\text{Sample Volume, mL}}$$

$$\text{Amps} = \frac{\text{Volts}}{\text{Ohms}}$$

$$\text{Area of Circle*} = (0.785)(\text{Diameter}^2)$$

$$\text{Area of Circle} = (3.14)(\text{Radius}^2)$$

$$\text{Area of Cone (lateral area)} = (3.14)(\text{Radius})\sqrt{\text{Radius}^2 + \text{Height}^2}$$

$$\text{Area of Cone (total surface area)} = (3.14)(\text{Radius})(\text{Radius} + \sqrt{\text{Radius}^2 + \text{Height}^2})$$

$$\text{Area of Cylinder (total exterior surface area)} = [\text{End \#1 SA}] + [\text{End \#2 SA}] + [(3.14)(\text{Diameter})(\text{Height or Depth})]$$
Where SA = surface area

$$\text{Area of Rectangle*} = (\text{Length})(\text{Width})$$

$$\text{Area of Right Triangle*} = \frac{(\text{Base})(\text{Height})}{2}$$

$$\text{Average (arithmetic mean)} = \frac{\text{Sum of All Terms}}{\text{Number of Terms}}$$

$$\text{Average (geometric mean)} = [(X_1)(X_2)(X_3)(X_4)(X_n)]^{1/n} \qquad \text{The } nth \text{ root of the product of } n \text{ numbers}$$

$$\text{Biochemical Oxygen Demand (seeded), mg/L} = \frac{[(\text{Initial DO, mg/L}) - (\text{Final DO, mg/L}) - (\text{Seed Correction, mg/L})][300\ \text{mL}]}{\text{Sample Volume, mL}}$$

$$\text{Biochemical Oxygen Demand (unseeded), mg/L} = \frac{[(\text{Initial DO, mg/L}) - (\text{Final DO, mg/L})][300\ \text{mL}]}{\text{Sample Volume, mL}}$$

$$\text{Blending or Three Normal Equation} = (C_1 \times V_1) + (C_2 \times V_2) = (C_3 \times V_3)$$
Where $V_1 + V_2 = V_3$; C = concentration, V = volume or flow; Concentration units must match; Volume units must match

$$\text{\# CFU/100 mL} = \frac{[(\text{\# of Colonies on Plate})(100)]}{\text{Sample Volume, mL}}$$

$$\text{Chemical Feed Pump Setting, \% Stroke} = \frac{\text{Desired Flow}}{\text{Maximum Flow}} \times 100\%$$

$$\text{Chemical Feed Pump Setting, mL/min} = \frac{(\text{Flow, mgd})(\text{Dose, mg/L})(3.785\ \text{L/gal})(1\ 000\ 000\ \text{gal/mil. gal})}{(\text{Feed Chemical Density, mg/mL})(\text{Active Chemical, \% express as a decimal})(1440\ \text{min/d})}$$

$$\text{Chemical Feed Pump Setting, mL/min} = \frac{(\text{Flow, m}^3\text{/d})(\text{Dose, mg/L})}{(\text{Feed Chemical Density, g/cm}^3)(\text{Active Chemical, \% express as a decimal})(1440\ \text{min/d})}$$

Circumference of Circle = (3.14)(Diameter)

$$\text{Composite Sample Single Portion} = \frac{(\text{Instantaneous Flow})(\text{Total Sample Volume})}{(\text{Number of Portions})(\text{Average Flow})}$$

$$\text{Cycle Time, min} = \frac{\text{Storage Volume, gal}}{(\text{Pump Capacity, gpm}) - (\text{Wet Well Inflow, gpm})}$$

$$\text{Cycle Time, min} = \frac{\text{Storage Volume, m}^3}{(\text{Pump Capacity, m}^3/\text{min}) - (\text{Wet Well Inflow, m}^3/\text{min})}$$

$$\text{Degrees Celsius} = \frac{(°\text{F} - 32)}{1.8}$$

Degrees Fahrenheit = (ºC)(1.8) + 32

$$\text{Detention Time} = \frac{\text{Volume}}{\text{Flow}} \qquad \textit{Units must be compatible}$$

Dilution or Two Normal Equation = $(C_1 \times V_1) = (C_2 \times V_2)$ *Where C = Concentration, V = volume or flow; Concentration units must match; Volume units must match*

Electromotive Force, V* = (Current, A)(Resistance, ohm - Ω)

$$\text{Feed Rate, lb/d*} = \frac{(\text{Dosage, mg/L})(\text{Flow, mgd})(8.34 \text{ lb/gal})}{\text{Purity, \% expressed as a decimal}}$$

$$\text{Feed Rate, kg/d*} = \frac{(\text{Dosage, mg/L})(\text{Flowrate, m}^3/\text{d})}{(\text{Purity, \% expressed as a decimal})(1000)}$$

$$\text{Filter Backwash Rate, gpm/sq ft} = \frac{\text{Flow, gpm}}{\text{Filter Area, sq ft}}$$

$$\text{Filter Backwash Rate, L/(m}^2\text{·s)} = \frac{\text{Flow, L/s}}{\text{Filter Area, m}^2}$$

$$\text{Filter Backwash Rise Rate, in./min} = \frac{(\text{Backwash Rate, gpm/sq ft})(12 \text{ in./ft})}{7.48 \text{ gal/cu ft}}$$

$$\text{Filter Backwash Rise Rate, cm/min} = \frac{\text{Water Rise, cm}}{\text{Time, min}}$$

$$\text{Filter Yield, lb/sq ft/hr} = \frac{(\text{Solids Loading, lb/d})(\text{Recovery, \% expressed as a decimal})}{(\text{Filter Operation, hr/d})(\text{Area, sq ft})}$$

$$\text{Filter Yield, kg/m}^2\text{·h} = \frac{(\text{Solids Concentration, \% expressed as a decimal})(\text{Sludge Feed Rate, L/h})(10)}{(\text{Surface Area of Filter, m}^2)}$$

Flowrate, cu ft/sec* = (Area, sq ft)(Velocity, ft/sec)

Flowrate, m³/sec* = (Area, m²)(Velocity, m/s)

$$\text{Food-to-Microorganism Ratio} = \frac{\text{BOD}_5, \text{ lb/d}}{\text{MLVSS, lb}}$$

$$\text{Food-to-Microorganism Ratio} = \frac{\text{BOD}_5, \text{ kg/d}}{\text{MLVSS, kg}}$$

Force, lb* = (Pressure, psi)(Area, sq in.)

Force, newtons* = (Pressure, Pa)(Area, m²)

Hardness, as mg CaCO₃/L = $\dfrac{\text{(Titrant Volume, mL)(1000)}}{\text{Sample Volume, mL}}$ *Only when the titration factor is 1.00 of ethylenediaminetetraacetic acid (EDTA)*

Horsepower, Brake, hp = $\dfrac{\text{(Flow, gpm)(Head, ft)}}{\text{(3960)(Pump Efficiency, \% expressed as a decimal)}}$

Horsepower, Brake, kW = $\dfrac{\text{(9.8)(Flow, m}^3\text{/s)(Head, m)}}{\text{(Pump Efficiency, \% expressed as a decimal)}}$

Horsepower, Motor, hp = $\dfrac{\text{(Flow, gpm)(Head, ft)}}{\text{(3960)(Pump Efficiency, \% expressed as a decimal)(Motor Efficiency, \% expressed as a decimal)}}$

Horsepower, Motor, kW = $\dfrac{\text{(9.8)(Flow, m}^3\text{/s)(Head, m)}}{\text{(Pump Efficiency, \% expressed as a decimal)(Motor Efficiency, \% expressed as a decimal)}}$

Horsepower, Water, hp = $\dfrac{\text{(Flow, gpm)(Head, ft)}}{3960}$

Horsepower, Water, kW = (9.8)(Flow, m³/s)(Head, m)

Hydraulic Loading Rate, gpd/sq ft = $\dfrac{\text{Total Flow Applied, gpd}}{\text{Area, sq ft}}$

Hydraulic Loading Rate, m³/(m²·d) = $\dfrac{\text{Total Flow Applied, m}^3\text{/d}}{\text{Area, m}^2}$

Loading Rate, lb/d* = (Flow, mgd)(Concentration, mg/L)(8.34 lb/gal)

Loading Rate, kg/d* = $\dfrac{\text{(Flow, m}^3\text{/d)(Concentration, mg/L)}}{1000}$

Mass, lb* = (Volume, mil. gal)(Concentration, mg/L)(8.34 lb/gal)

Mass, kg* = $\dfrac{\text{(Volume, m}^3\text{)(Concentration, mg/L)}}{1000}$

Mean Cell Residence Time or Solids Retention Time, days = $\dfrac{\text{(Aeration Tank TSS, lb)} + \text{(Clarifier TSS, lb)}}{\text{(TSS Wasted, lb/d)} + \text{(Effluent TSS, lb/d)}}$

Milliequivalent = (mL)(Normality)

Molarity = $\dfrac{\text{Moles of Solute}}{\text{Liters of Solution}}$

Motor Efficiency, % = $\dfrac{\text{Brake hp}}{\text{Motor hp}} \times 100\%$

Normality = $\dfrac{\text{Number of Equivalent Weights of Solute}}{\text{Liters of Solution}}$

Number of Equivalent Weights = $\dfrac{\text{Total Weight}}{\text{Equivalent Weight}}$

$$\text{Number of Moles} = \frac{\text{Total Weight}}{\text{Molecular Weight}}$$

$$\text{Organic Loading Rate-RBC, lb SBOD}_5/1000 \text{ sq ft/d} = \frac{\text{Organic Load, lb SBOD}_5/\text{d}}{\text{Surface Area of Media, 1000 sq ft}}$$

$$\text{Organic Loading Rate-RBC, kg SBOD}_5/\text{m}^2\cdot\text{d} = \frac{\text{Organic Load, kg SBOD}_5/\text{d}}{\text{Surface Area of Media, m}^2}$$

$$\text{Organic Loading Rate-Trickling Filter, lb BOD}_5/1000 \text{ cu ft/d} = \frac{\text{Organic Load, lb BOD}_5/\text{d}}{\text{Volume, 1000 cu ft}}$$

$$\text{Organic Loading Rate-Trickling Filter, kg/m}^3\cdot\text{d} = \frac{\text{Organic Load, kg BOD}_5/\text{d}}{\text{Volume, m}^3}$$

$$\text{Oxygen Uptake Rate or Oxygen Consumption Rate, mg/L}\cdot\text{min} = \frac{\text{Oxygen Usage, mg/L}}{\text{Time, min}}$$

$$\text{Population Equivalent, Organic} = \frac{(\text{Flow, mgd})(\text{BOD, mg/L})(8.34 \text{ lb/gal})}{0.17 \text{ lb BOD/d/person}}$$

$$\text{Population Equivalent, Organic} = \frac{(\text{Flow, m}^3/\text{d})(\text{BOD, mg/L})}{(1000)(0.077 \text{ kg BOD/d}\cdot\text{person})}$$

$$\text{Power, kW} = \frac{(\text{Flow, L/s})(\text{Head, m})(9.8)}{1000}$$

$$\text{Recirculation Ratio-Trickling Filter} = \frac{\text{Recirculated Flow}}{\text{Primary Effluent Flow}}$$

$$\text{Reduction of Volatile Solids, \%} = \left(\frac{\text{VS in} - \text{VS out}}{\text{VS in} - (\text{VS in} \times \text{VS out})}\right) \times 100\% \qquad \textit{All information (In and Out) must be in decimal form}$$

$$\text{Removal, \%} = \left(\frac{\text{In} - \text{Out}}{\text{In}}\right) \times 100\%$$

$$\text{Return Rate, \%} = \frac{\text{Return Flowrate}}{\text{Influent Flowrate}} \times 100\%$$

$$\text{Return Sludge Rate-Solids Balance, mgd} = \frac{(\text{MLSS, mg/L})(\text{Flowrate, mgd})}{(\text{RAS Suspended Solids, mg/L}) - (\text{MLSS, mg/L})}$$

$$\text{Slope, \%} = \frac{\text{Drop or Rise}}{\text{Distance}} \times 100\%$$

$$\text{Sludge Density Index} = \frac{100}{\text{SVI}}$$

$$\text{Sludge Volume Index, mL/g} = \frac{(\text{SSV}_{30}, \text{mL/L})(1000 \text{ mg/g})}{\text{MLSS, mg/L}}$$

$$\text{Solids, mg/L} = \frac{(\text{Dry Solids, g})(1\,000\,000)}{\text{Sample Volume, mL}}$$

$$\text{Solids Capture, \% (Centrifuges)} = \left[\frac{\text{Cake TS, \%}}{\text{Feed Sludge TS, \%}}\right] \times \left[\frac{(\text{Feed Sludge TS, \%}) - (\text{Centrate TSS, \%})}{(\text{Cake TS, \%}) - (\text{Centrate TSS, \%})}\right] \times 100\%$$

$$\text{Solids Concentration, mg/L} = \frac{\text{Weight, mg}}{\text{Volume, L}}$$

$$\text{Solids Loading Rate, lb/sq ft/d} = \frac{\text{Solids Applied, lb/d}}{\text{Surface Area, sq ft}}$$

$$\text{Solids Loading Rate, kg/m}^2\text{·d} = \frac{\text{Solids Applied, kg/d}}{\text{Surface Area, m}^2}$$

Solids Retention Time: *see Mean Cell Residence Time*

$$\text{Specific Gravity} = \frac{\text{Specific Weight of Substance, lb/gal}}{8.34 \text{ lb/gal}}$$

$$\text{Specific Gravity} = \frac{\text{Specific Weight of Substance, kg/L}}{1.0 \text{ kg/L}}$$

$$\text{Specific Oxygen Uptake Rate or Respiration Rate, (mg/g)/h} = \frac{(\text{OUR, mg/L·min})(60 \text{ min})}{(\text{MLVSS, g/L})(1 \text{ h})}$$

$$\text{Surface Loading Rate or Surface Overflow Rate, gpd/sq ft} = \frac{\text{Flow, gpd}}{\text{Area, sq ft}}$$

$$\text{Surface Loading Rate or Surface Overflow Rate, L/m}^2\text{·d} = \frac{\text{Flow, L/d}}{\text{Area, m}^2}$$

$$\text{Total Solids, \%} = \frac{(\text{Dried Weight, g}) - (\text{Tare Weight, g})}{(\text{Wet Weight, g}) - (\text{Tare Weight, g})} \times 100\%$$

$$\text{Velocity, ft/sec} = \frac{\text{Flowrate, cu ft/s}}{\text{Area, sq ft}}$$

$$\text{Velocity, ft/sec} = \frac{\text{Distance, ft}}{\text{Time, sec}}$$

$$\text{Velocity, m/s} = \frac{\text{Flowrate, m}^3\text{/s}}{\text{Area, m}^2}$$

$$\text{Velocity, m/s} = \frac{\text{Distance, m}}{\text{Time, s}}$$

$$\text{Volatile Solids, \%} = \left[\frac{(\text{Dry Solids, g}) - (\text{Fixed Solids, g})}{(\text{Dry Solids, g})} \right] \times 100\%$$

Volume of Cone* $= (1/3)(0.785)(\text{Diameter}^2)(\text{Height})$

Volume of Cylinder* $= (0.785)(\text{Diameter}^2)(\text{Height})$

Volume of Rectangular Tank* $= (\text{Length})(\text{Width})(\text{Height})$

$$\text{Water Use, gpcd} = \frac{\text{Volume of Water Produced, gpd}}{\text{Population}}$$

$$\text{Water Use, L/cap·d} = \frac{\text{Volume of Water Produced, L/d}}{\text{Population}}$$

Watts (AC circuit) = (Volts)(Amps)(Power Factor)

Watts (DC circuit) = (Volts)(Amps)

$$\text{Weir Overflow Rate, gpd/ft} = \frac{\text{Flow, gpd}}{\text{Weir Length, ft}}$$

$$\text{Weir Overflow Rate, L/m·d} = \frac{\text{Flow, L/d}}{\text{Weir Length, m}}$$

$$\text{Wire-to-Water Efficiency, \%} = \frac{\text{Water hp}}{\text{Motor hp}} \times 100\%$$

$$\text{Wire-to-Water Efficiency, \%} = \frac{(\text{Flow, gpm})(\text{Total Dynamic Head, ft})(0.746 \text{ kW/hp})(100\%)}{(3960)(\text{Electrical Demand, kW})}$$

Pie Wheels

- To find the quantity above the horizontal line: multiply the pie wedges below the line together.
- To solve for one of the pie wedges below the horizontal line: cover that pie wedge, then divide the remaining pie wedge(s) into the quantity above the horizontal line.
- Given units must match the units shown in the pie wheel.
- When US and metric units or values differ, the metric is shown in parentheses, e.g. (m^2).

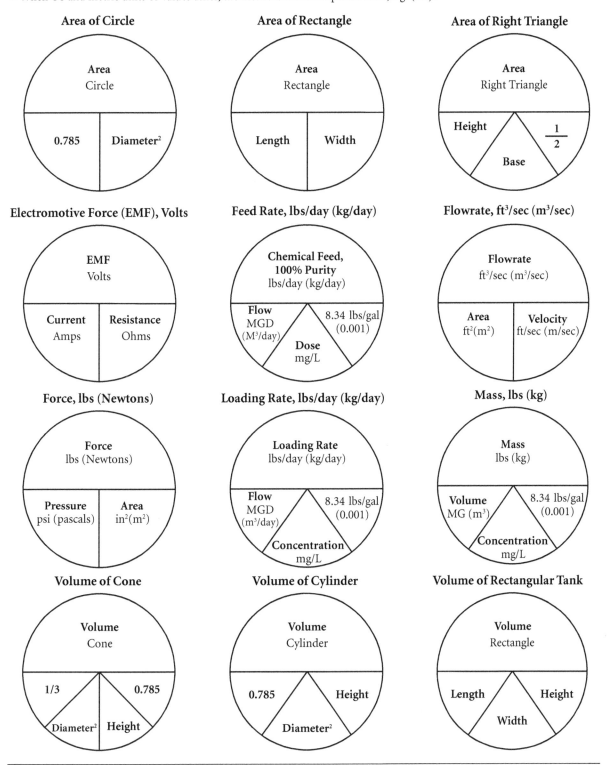

Conversion Factors

1 ac = 4046.9 m^2 or 43 560 sq ft

1 ac ft of water = 326 000 gal

1 atm = 33.9 ft of water

    = 10.3 m of water

    = 14.7 psi

    = 101.3 kPa

1 cfs = 0.646 mgd

    = 448.8 gpm

1 cu ft of water = 7.48 gal

    = 62.4 lb

1 ft = 0.305 m

1 ft H$_2$O = 0.433 psi

1 gal (US) = 3.79 L

    = 8.34 lb of water

1 gr/gal (US) = 17.1 mg/L

1 ha = 10 000 m^2

1 hp = 0.746 kW

    = 746 W

    = 33 000 ft lb/min

1 in. = 25.4 mm or 2.54 cm

1 L/s = 0.0864 ML/d

1 lb = 0.454 kg

1 m of water = 9.8 kPa

1 m^2 = 1.19 sq yd

1 m^3 = 1000 kg

    = 1000 L

    = 264 gal

1 metric ton = 2205 lb

1 mile = 5280 ft

1 mgd = 694 gpm

    = 1.55 cfs

    = 3.785 ML/d

Population equivalent (PE), hydraulic = 378.5 L/cap·d

    = 100 gpd/cap

PE, organic = 0.077 kg BOD/cap·d

    = 0.17 lb BOD/cap/d

1 psi = 2.31 ft of water

    = 6.89 kPa

1 ton = 2000 lb

1% = 10 000 mg/L

π or pi = 3.14

CHAPTER 1
Introduction to Solids Handling

Solids Characteristics

1. Natural treatment systems typically treat liquids and solids in the same processes.
    - ☐ True
    - ☐ False

2. Scum removed from the surface of a primary clarifier may be combined with primary sludge removed from the bottom of the clarifier before being transferred to the solids handling site of the WRRF.
    - ☐ True
    - ☐ False

3. Secondary sludge tends to have higher total suspended solids concentrations than primary sludge.
    - ☐ True
    - ☐ False

4. This term may be used to describe solid material removed from wastewater during treatment.
    - a. Scum
    - b. Residuals
    - c. Clink
    - d. Volatiles

5. Chemical sludge may be produced at all of the following locations EXCEPT
    - a. Primary clarifier
    - b. Secondary clarifier
    - c. Tertiary filter
    - d. Chlorine contact chamber

6. An example of biological, secondary sludge would be
    - a. Solids removed during primary clarification
    - b. Solids generated during phosphorus precipitation
    - c. Solids removed from the activated sludge process
    - d. Solids removed during screening and grit removal

7. This type of residual is typically sent to a landfill.
    - a. Grit
    - b. Primary sludge
    - c. Secondary sludge
    - d. Chemical sludge

8. Primary sludge consists primarily of
    - a. Sand, gravel, fruit rinds, and seeds
    - b. Fibrous materials, cellulose, and organic and inorganic solids
    - c. Microorganisms produced in the treatment process
    - d. Precipitated iron or aluminum compounds

9. This type of sludge tends to have a higher percentage of volatile solids than other types of sludge.
    a. Grit
    b. Secondary
    c. Chemical
    d. Dewatered

10. Chemical sludge generated by adding ferric for odor control, to enhance clarifier performance, or precipitate phosphorus tends to contain more _____ than other types of sludge.
    a. Inert material
    b. Microorganisms
    c. Cellulose
    d. Pathogens

11. Match the return stream to the process that produced it.

1.	Filtrate	a.	Rotary drum thickener
2.	Centrate	b.	Digesters and gravity thickeners
3.	Drummate	c.	Belt filter press or plate-and-frame press
4.	Supernatant	d.	Tertiary filters
5.	Backwash	e.	Centrifuge

12. Return flows from solids handling processes must be managed to
    a. Maximize return flows in the middle of the day.
    b. Reduce opportunities for ammonia pass-through.
    c. Minimize capture of fine solids.
    d. Ensure smooth operation of solids handling equipment.

Solids Handling

1. Once solids have been separated from the treated wastewater, treatment is complete.
    ☐ True
    ☐ False

2. Some sources of contaminants in sludge include homes, businesses, industrial users, and the drinking water supply.
    ☐ True
    ☐ False

3. Contaminant concentrations in sludge are often high enough to be harmful to operators.
    ☐ True
    ☐ False

4. Trace amounts of heavy metals in sludge come from all of the following sources EXCEPT
    a. Homes and offices
    b. Settleable solids
    c. Drinking water supply
    d. Dissolved solids

5. An example of a vector would be
    a. Virus
    b. Bacteria
    c. Insect
    d. Parasite

6. Another term for land application of biosolids is
    a. Beneficial reuse
    b. Surface disposal
    c. Landfilling
    d. Discing or plowing

7. Before biosolids can be land applied, they must meet the requirements of this important regulation.

    a.   40 CFR Part 258

    b.   Resource Conservation and Recovery Act (RCRA)

    c.   40 CFR Part 503

    d.   Occupational Safety and Health Act (OSHA)

8. Without looking back in the text, list the goals of solids handling.

    a.   _____

    b.   _____

    c.   _____

    d.   _____

    e.   _____

Reducing Volume

1. Primary and secondary sludge contain more solids than water.

    ☐ True

    ☐ False

2. One reason for increasing the solids concentrations in sludge is to reduce the size of downstream processes.

    ☐ True

    ☐ False

3. It is possible to remove all of the water from sludge with pressure.

    ☐ True

    ☐ False

4. This influent parameter may be used for estimating sludge production from a primary clarifier.

    a. BOD_5

    b.   TVS

    c.   TSS

    d.   Flow

5. Given the following information, find the mass of solids entering a primary clarifier. Influent flow = 18 925 m³/d (5 mgd), Influent BOD_5 = 250 mg/L, Influent TSS = 280 mg/L. The primary clarifier removes 35% of the influent TSS.

    a.   4242 kg/d (9350 lb/d)

    b.   4731 kg/d (10 425 lb/d)

    c.   4751 kg/d (10 472 lb/d)

    d.   5299 kg/d (11 676 lb/d)

6. The influent flow to a WRRF is 11 355 m³/d (3 mgd) and it contains 315 mg/L of TSS. If the primary clarifier removes 40% of the influent TSS, how many kilograms (pounds) of primary sludge will be generated each day?

    a.   1431 kg/d (3152 lb/d)

    b.   1924 kg/d (4241 lb/d)

    c.   2146 kg/d (4729 lb/d)

    d.   3577 kg/d (7881 lb/d)

7. Sludge from a primary clarifier is 5.3% total solids. What is this in milligrams per liter (mg/L)?

    a.   530 mg/L

    b.   5300 mg/L

    c.   53 000 mg/L

    d.   530 000 mg/L

8. A secondary clarifier produces 190 m³/d (50 000 gpd) of waste activated sludge (WAS). If the concentration is increased from 7000 mg/L to 3%, how many gallons of sludge will remain?
   a. 35.2 m³ (9325 gal)
   b. 44.3 m³ (11 667 gal)
   c. 93.3 m³ (25 428 gal)
   d. 117.8 m³ (31 113 gal)

9. A primary clarifier produces 1854.65 kg/d (4086.6 lb/d) of total solids. Find the volume of sludge produced in m³/d (gpd) if the average sludge concentration is 8%.
   a. 23.18 m³/d (6125 gpd)
   b. 25.85 m³/d (6829 gpd)
   c. 231.8 m³/d (61 250 gpd)
   d. 2585.0 m³/d (68 290 gpd)

10. This type of water is easiest to remove from sludge and biosolids.
    a. Free
    b. Interstitial
    c. Vicinal
    d. Water of hydration

11. In addition to particle size, this parameter has the biggest effect on removing water from sludge.
    a. pH
    b. Oxygen content
    c. Percentage volatile solids
    d. Iron content

12. Which of the following types of solids can be thickened or dewatered most easily?
    a. Sand
    b. Fine mud
    c. Primary sludge
    d. Secondary sludge

13. Mechanical dewatering methods such as belt filter presses and centrifuges are unable to achieve biosolids concentrations greater than 60% for this reason.
    a. The amount of polymer required is harmful to the equipment.
    b. Vicinal water cannot be removed mechanically.
    c. Operating costs would be excessive.
    d. Polymer added for dewatering absorbs water.

14. Smaller sludge particles are more difficult to dewater than large sludge particles because
    a. They have less surface area per volume.
    b. They contain more free water.
    c. They have more surface area per volume.
    d. They contain less free water.

Sludge Conditioning

1. Sludge conditioning adds chemicals to sludge or biosolids to allow water to drain away easier.
   ☐ True
   ☐ False

2. A disadvantage of organic conditioners compared to inorganic conditioners is that they add significant mass to the sludge cake, potentially increasing handling costs.
   ☐ True
   ☐ False

3. Viscosity may be used to estimate polymer content.
    - ☐ True
    - ☐ False

4. Diluted polymer is not stable and may become ineffective if allowed to age in the mix tank too long.
    - ☐ True
    - ☐ False

5. Polymer spills should be cleaned up immediately to prevent slippery, hazardous conditions.
    - ☐ True
    - ☐ False

6. The conditioning process is used to prepare sludge or biosolids for more efficient
    - a. Thickening and disinfection
    - b. Stabilization and dilution
    - c. Thickening and dewatering
    - d. Neutralization and dewatering

7. This term describes the initial contact between a conditioning chemical and sludge particles to form first-stage floc particles.
    - a. Coagulation
    - b. Flocculation
    - c. Sedimentation
    - d. Filtration

8. Organic chemicals that are commonly used for conditioning include
    - a. Alum
    - b. Lime
    - c. Polymers
    - d. Ferric chloride

9. The most common polymer type used to condition wastewater treatment solids is
    - a. Anionic
    - b. Cationic
    - c. Non-ionic
    - d. Insoluble

10. This sludge conditioning chemical consumes alkalinity when added to water and may lower the pH of the conditioned sludge.
    - a. Alum
    - b. Lime
    - c. Ferric chloride
    - d. Calcium oxide

11. Lime is sometimes added to sludge along with ferric chloride for this reason.
    - a. To fully activate the ferric chloride
    - b. Prevent an unacceptable drop in pH
    - c. Decrease sludge production
    - d. Add ballast for improved settling

12. Which type of sludge may require a non-ionic or anionic polymer?
    - a. Primary sludge
    - b. Secondary sludge
    - c. Chemical sludge
    - d. Septic sludge

13. A WRRF must meet new discharge permit limits for phosphorus. Phosphorus will be precipitated with ferric chloride. Which of the following statements is NOT true?
    a. Operators may notice a reduction in odors.
    b. Sludge dewatering should improve.
    c. Lime must be added to prevent pH drop.
    d. A cationic polymer will work best for dewatering.

14. Polymer charge density is a measure of
    a. Coulomb count
    b. Amp draw
    c. Charges per length
    d. Molecular weight

15. Polymers are preferred over inorganic chemicals for sludge dewatering because
    a. Inorganic chemicals increase sludge mass.
    b. Polymers are more soluble in oily sludge.
    c. Inorganic chemicals decrease handling costs.
    d. Polymers require prewetting and aging.

16. Caking of dry polymer can be prevented by
    a. Wrapping opened totes in brown paper
    b. Storing containers in a cool, dry area
    c. Misting to maintain moisture levels
    d. Coating storage bins with oil

17. One advantage of using dry polyacrylamide polymers (PAM) over other forms is it
    a. Does not require make-down
    b. Contains 100% active product
    c. May be stored in a moist environment
    d. Offers lowest shipping cost per unit

18. During polymer make-down, the polymer must _____ to ensure the polymer is in the right form to flocculate sludge particles.
    a. Age
    b. Settle
    c. Aerate
    d. Cure

19. This type of polymer does not require make-down or aging.
    a. Dry polymer
    b. Emulsion polymer
    c. Mannich polymer
    d. Salted polymer

Laboratory Testing

1. When conducting jar tests, operators should use the largest dose of chemical needed to get the absolute best performance.
    ☐ True
    ☐ False

2. Operators should try to achieve the driest cake possible when dewatering.
    ☐ True
    ☐ False

3. A metric ton is equivalent to 2000 lb.
   - ☐ True
   - ☐ False

4. _____ tests are commonly used to screen conditioning agents, especially when a wide variety of potentially effective products is available.
   a. Jar
   b. Water absorption
   c. Solids capture
   d. Volatile solids

5. Jar testing should be done with this type of container to maximize turbulence and mixing.
   a. Round
   b. Square
   c. Plastic
   d. Glass

6. When sludge has been properly conditioned, thickening and dewatering processes are capable of capturing _____ percent or more of sludge particles.
   a. 50
   b. 75
   c. 95
   d. 100

7. Each of these parameters is used to evaluate a chemical conditioner's performance in jar testing EXCEPT
   a. Floc formation
   b. Supernatant clarity
   c. Filtrate volume
   d. Chemical dose

8. Four jar tests were performed on secondary sludge using different doses of polymer. The lowest dose of polymer produced weak, small flocs and cloudy supernatant. The next-to-lowest dose produced strong floc, but the supernatant remained cloudy. The third test produced strong floc and relatively clear supernatant. The fourth test had almost identical results to the third test, but crystal-clear supernatant. The polymer dosages used for the test were 30 mg/L, 60 mg/L, 90 mg/L, and 150 mg/L. Which dose should be selected for use in the WRRF process?
   a. 30 mg/L
   b. 60 mg/L
   c. 90 mg/L
   d. 150 mg/L

9. Jar testing has determined that the optimum dose for 600 mL of sludge is 5 mL of a 20 000 mg/L polymer solution. What is the concentration of polymer in the jar test?
   a. 122 mg/L
   b. 167 mg/L
   c. 184 mg/L
   d. 221 mg/L

10. Polymer is fed at a dose of 75 mg/L. Find the polymer feed rate in kg/d (lb/d) if the sludge feed rate is 5 L/s (79.3 gpm).
    a. 32.4 kg/d (71.4 lb/d)
    b. 48.6 kg/d (107 lb/d)
    c. 64.8 kg/d (143 lb/d)
    d. 81.0 kg/d (179 lb/d)

Reduction Requirements

1. In endogenous respiration, bacteria convert VSS to methane and carbon dioxide.
   - ☐ True
   - ☐ False

2. Sludge digestion can reduce the VSS in sludge by as much as 50%.
   - ☐ True
   - ☐ False

3. Drying sludge to less than 10% water is one method of inactivating or killing pathogens during sludge stabilization.
   - ☐ True
   - ☐ False

4. Before biosolids may be land applied, they must meet criteria for pathogen reduction, vector attraction reduction, and contaminants.
   - ☐ True
   - ☐ False

5. When aerobic and anaerobic digesters are operated at cooler temperatures, what must be true?
   a. Gas production increases
   b. Digestion takes more time
   c. Storage space is decreased
   d. More inert TSS is broken down

6. During aerobic digestion, volatile solids are broken down to form
   a. Methane and carbon dioxide
   b. Volatile fatty acids and water
   c. Carbon dioxide and water
   d. Ammonia and methane

7. Anaerobic digesters produce this gas, which may be burned to generate electricity.
   a. Methane
   b. Carbon dioxide
   c. Propane
   d. Nitrogen

8. This class of biosolids can be made available to the public.
   a. Class A
   b. Class B
   c. Class C
   d. Class D

9. The SOUR test is used to demonstrate
   a. Activated sludge process efficiency
   b. Pathogens have been reduced to acceptable levels
   c. Contaminants are below allowable limits
   d. Vector attraction reduction requirements have been met

10. Raw sludge entering a digester contains 83% volatile solids. The finished biosolids contain 71% volatile solids. Find the percent reduction of volatile solids.
    a. 12.0%
    b. 14.4%
    c. 49.8%
    d. 72.0%

CHAPTER 2
Thickening

Purpose and Function of Thickening

1. Thickening commonly occurs before what process?
   a. Conditioning
   b. Grit removal
   c. Disposal
   d. Stabilization

2. Where is the water removed during thickening sent?
   a. Dewatering
   b. Final discharge
   c. Headworks
   d. Digestion

3. The main benefit of thickening is an increase in
   a. Water
   b. Heating
   c. Volatile content
   d. Solids

Gravity Thickening

1. A gravity thickener is used only to thicken WAS.
   ☐ True
   ☐ False

2. A high torque reading on the drive unit of the collector mechanism is caused by
   a. Low sludge blanket
   b. High hydraulic flowrate through the gravity thickener
   c. The collector is turned off
   d. The sludge blanket in the gravity thickener is deep and heavy

3. A typical circular gravity thickener has a flat bottom.
   ☐ True
   ☐ False

4. Label the parts in the following picture:

(Reprinted with permission of Lyle Leubner, North Little Rock Wastewater Facility)

a. _____

b. _____

c. _____

d. _____

5. A gravity thickener that receives only primary sludge typically thickens the sludge to what concentration?

a. 0.5%–1%

b. 1%–5%

c. 5%–10%

d. 10%–15%

6. A gravity thickened sludge concentration of 8% to 10% is typically pumped by a(n) _____?

a. Progressing cavity type pump

b. Centrifugal pump

c. Diaphragm pump

d. Air lift pump

7. The bottom of a gravity thickener usually has a slope of 2 : 12 to 3 : 12.

☐ True

☐ False

8. Rat-holing is when water is drawn down through a hole in the sludge blanket because the solids removal rate is too high.

☐ True

☐ False

9. The typical detention time of a primary fed gravity thickener is 18 to 30 hours.

☐ True

☐ False

Dissolved Air Flotation Thickening

1. Large rectangular DAFT units typically have both a top and bottom skimmer mechanism.

☐ True

☐ False

2. The pressurized or recycle flowrate of a DAFT is typically set at what percentage of the incoming flow?

a. 0%–10%

b. 50%–60%

c. 75%–100%

d. 100%–200%

3. A DAFT can operate with or without polymer.
   ☐ True
   ☐ False

4. An operator conducts a rise test in a graduated cylinder and the test shows that there is a very thick solids layer at the top. What does this indicate?
   a. The DAFT is underloaded.
   b. The DAFT is overloaded.
   c. The top skimmer mechanism is operating too fast.
   d. The top skimmer mechanism is operating too slowly.

5. Dissolved air flotation thickeners are typically designed to have a float concentration of _____?
   a. 2%
   b. 4%
   c. 6%
   d. 8%

6. The float layer of a DAFT is typically operated at more than 915 mm (36 in.) in depth.
   ☐ True
   ☐ False

7. Which of the following is a potential cause of the DAFT effluent (subnatant) suspended solids being too high?
   a. Polymer dosage is too high
   b. DAFT unit is underloaded
   c. Skimmer operating too slowly
   d. Too high air-to-solids ratio

8. One reason for low dissolved air in the DAFT is that the compressed air system is set too high.
   ☐ True
   ☐ False

9. The rise rate being too slow typically indicates that the DAFT unit is overloaded.
   ☐ True
   ☐ False

10. The float solids from the DAFT are sent to _____.
    a. Disinfection
    b. Headworks
    c. Digestion
    d. Clarification

Gravity Belt Thickeners

1. Polymer is used with a GBT to _____.
   a. Reduce solids loading
   b. Increase solids loading
   c. Reduce capture rate
   d. Increase capture rate

2. The typical capture rate of a GBT is _____.
   a. 65%
   b. 75%
   c. 85%
   d. 95%

3. Chicanes on the GBT are used to _____.
    a. Increase belt detention time
    b. Move solids into rows
    c. Mix in polymer solution
    d. Move the solids to the center

4. The filtrate from the GBT should be black in color.
    ☐ True
    ☐ False

5. The spray water for the GBT is used to _____.
    a. Help condition the polymer
    b. Help release the sludge from the belt
    c. Clean the chicanes
    d. Clean the flocculation tank

6. The polymer solution is typically injected to the feed sludge through an injection ring.
    ☐ True
    ☐ False

Centrifuge Thickening

1. Centrifuge thickening units can be operated with or without polymer.
    ☐ True
    ☐ False

2. What is the purpose of the weir (dam) at the effluent end of the centrifuge?
    a. Adjust the differential speed
    b. Adjust the torque
    c. Adjust the pool depth
    d. Adjust the polymer feed rate

3. Observing the color of the centrate is an important process control parameter for operators.
    ☐ True
    ☐ False

4. Thickened sludge concentration can be increased or decreased by adjusting the pool depth.
    ☐ True
    ☐ False

5. The bowl and the scroll operate at the same speed in a centrifuge.
    ☐ True
    ☐ False

6. Centrifuges used for thickening sludges are controlled to produce a solids concentration range of _____.
    a. 1%–2%
    b. 5%–6%
    c. 10%–11%
    d. 15%–16%

7. The target TSS rate in the centrate is _____.
    a. Less than 500 mg/L
    b. Less than 1000 mg/L
    c. Less than 1500 mg/L
    d. Less than 2000 mg/L

Rotary Drum Thickeners

1. Rotary drum thickeners require polymer to operate properly.
   - ☐ True
   - ☐ False

2. The overall average range of thickened sludge concentration from a rotary drum thickener is _____?
   - a. 1%–7%
   - b. 5%–12%
   - c. 9%–16%
   - d. 12%–19%

3. The typical capture rate of a rotary drum thickener that receives WAS is _____?
   - a. 63%–69%
   - b. 73%–79%
   - c. 83%–89%
   - d. 93%–99%

4. A rotary drum thickener typically uses _____ m³/h (gpm) of wash water?
   - a. 5–23 m³/h (20–100 gpm)
   - b. 10–35 m³/h (45–155 gpm)
   - c. 15–45 m³/h (65–200 gpm)
   - d. 20–50 m³/h (90–220 gpm)

5. Increasing the drum speed of a rotary drum thickener will _____.
   - a. Increase the thickened sludge concentration
   - b. Decrease the thickened sludge concentration
   - c. Increase the sludge throughput
   - d. Decrease the sludge throughput

6. A rotary drum thickener uses paddles or a screw conveyor inside of the drum to move the thickened sludge to the effluent end.
   - ☐ True
   - ☐ False

7. There is good flocculation in the mixing tank; however, the thickened sludge slurry is leaving the rotary drum thickener sloppily. What is a probable cause?
   - a. The polymer dose is too high.
   - b. The polymer dose is too low.
   - c. The drum is clogged.
   - d. The drum is clear of debris.

8. Facility effluent can be used as a source of water for the wash water system in a rotary drum thickener.
   - ☐ True
   - ☐ False

Polymer Systems

1. The purpose for using polymer during thickening and dewatering is to _____.
   - a. Reduce the amount of solids
   - b. Reduce the operating costs
   - c. Improve the separation in thickening and dewatering
   - d. Improve the methane production in digestion processes

2. Polymers act as _____ in WRRF sludges.
    a. Coagulants
    b. Disinfectants
    c. Absorbents
    d. Precipitants

3. _____ is used to determine the optimal polymer dose to use to condition a specific sludge.
    a. Computational fluid dynamics modeling
    b. An optimal polymer dose calculator
    c. Jar testing
    d. Beaker batching

4. Sludge particle _____ has the most significant effect on which type of polymer will be effective at promoting coagulation.
    a. Mineral composition
    b. Shape
    c. Electrical charge
    d. Density

5. A _____ polymer would most likely be used to condition WAS.
    a. High charge density, cationic
    b. High charge density, anionic
    c. Low charge density, nonionic
    d. Low charge density, anionic

6. Incorrect _____ of dry polymer can cause "fish eyes" to form in the polymer solution.
    a. Storage
    b. Initial wetting
    c. Aging
    d. Dilution

7. Number the following dry polymer makeup steps in the proper chronological order to ensure that effective polymer solution is created:
    _____ Aging
    _____ Injection to process stream
    _____ Low humidity storage
    _____ Prewetting
    _____ Low energy mixing
    _____ High energy mixing

8. _____ is needed to "break the emulsion" on an emulsion polymer makeup system.
    a. High temperature
    b. Long aging
    c. High energy mixing
    d. Gentle agitation

9. A component on most emulsion polymer makeup systems that required frequent cleaning is the _____.
    a. Polymer solution tank inlet filter
    b. Motor cooling jacket
    c. Bulk storage tank
    d. Neat polymer check valve

Safety Considerations

1. All lubrication and maintenance of thickening units should be done when the equipment is in operation.
    - ☐ True
    - ☐ False

2. Proper respiratory equipment should be used when working with a dry polymer system.
    - ☐ True
    - ☐ False

3. Facility effluent is added to a gravity thickener in some cases to _____.
    - a. Increase solids loading
    - b. Decrease solids loading
    - c. Increase DO
    - d. Decrease DO

4. Thickened equipment that is installed with a cover is maintained at a positive pressure inside the cover to prevent odors.
    - ☐ True
    - ☐ False

5. The recycle of filtrate or centrate that is dirty reduces the capacity of the WRRF.
    - ☐ True
    - ☐ False

CHAPTER 3
Aerobic Digestion

Biological Treatment Fundamentals Review

1. Heterotrophic bacteria obtain their carbon and energy from _____, whereas nitrifying bacteria obtain their carbon from _____.
   a. Alkalinity, ammonia
   b. BOD, ammonia
   c. Ammonia, alkalinity
   d. BOD, alkalinity

2. Denitrification converts _____ to _____.
   a. Nitrite and nitrate to nitrogen gas
   b. Nitrogen gas to ammonia
   c. Ammonia to nitrite and nitrate
   d. Ammonia to nitrogen gas

3. For denitrification to take place, the following condition must be met:
   a. BOD depleted
   b. All ammonia converted to nitrate
   c. DO no longer available
   d. HDT less than 20 minutes

4. Nitrification consumes alkalinity, which can cause
   a. Nitrite to accumulate
   b. Endogenous respiration
   c. Phosphorus release
   d. pH to decrease

5. Most bacteria in biological treatment processes grow and reproduce quickly when
   a. pH is near neutral and water is warm
   b. pH is greater than 7 and water is cold
   c. pH is less than 7 and water is warm
   d. pH is near neutral and water is cold

Theory of Operation

1. Aerobic digestion reduces the total mass of sludge.
   ☐ True
   ☐ False

2. Aerobic digesters contain a unique mix of microorganisms not found in other treatment processes.
   ☐ True
   ☐ False

3. The total solids concentration in a well-functioning digester tends to decrease over time unless the operator takes some action to thicken the solids.
   ☐ True
   ☐ False

4. Nitrifying bacteria that end up in the digester undergo endogenous respiration just like all of the other microorganisms.
   ☐ True
   ☐ False

5. Digesters that receive a mixture of primary and secondary sludge require more oxygen than digesters receiving only secondary sludge.
   ☐ True
   ☐ False

6. Modern WRRFs pair primary clarifiers with aerobic digesters.
   ☐ True
   ☐ False

7. Goals of aerobic digestion include all of the following EXCEPT
   a. Reduce number of pathogens
   b. Reduce vector attraction
   c. Reduce oxygen production
   d. Reduce total mass of solids

8. Which of the following statements regarding aerobic digesters is NOT true?
   a. Aerobic digesters are always operated with a DO greater than 1 mg/L.
   b. Aerobic digesters may be operated with or without sludge pre-thickening.
   c. Aerobic digesters may be decanted to remove excess water.
   d. Aerobic digesters convert VS to carbon dioxide and water.

9. Which test method should be used to determine the concentration of organic solids in a digester if the estimated TS concentration is 5%?
   a. TSS
   b. Total VSS
   c. TS
   d. Total volatile solids

10. These two nutrients are released through endogenous respiration.
    a. Nitrogen and phosphorus
    b. Carbon dioxide and water
    c. Alkalinity and carbon dioxide
    d. Phosphorus and manganese

11. Endogenous respiration releases ammonia, which combines with dissolved carbon dioxide. This causes
    a. Alkalinity to decrease
    b. Phosphorus to precipitate
    c. Alkalinity to increase
    d. Phosphorus to sublimate

12. Inorganic solids entering the digester
    a. Are converted into new biomass
    b. Pass through the digester unchanged
    c. Are a source of BOD for the microorganisms
    d. May be partially biodegradable

13. In most aerobic digesters, ammonia
    a. Accumulates to toxic concentrations
    b. Is converted to nitrite and nitrate
    c. Combines with phosphorus as struvite
    d. Decreases under anoxic conditions

14. Nitrification in digesters can be inhibited or slowed down by
    a. Maintaining SRTs longer than 40 days
    b. Operating with a high pH level
    c. Limiting the amount of oxygen available
    d. Using covers to increase water temperature

15. How much oxygen is needed to destroy 1 kg of VS and convert released ammonia to nitrate?
    a. 0.5 kg
    b. 1.0 kg
    c. 1.5 kg
    d. 2.0 kg

16. A digester receiving both primary and secondary sludge will
    a. Consume more oxygen than a digester receiving only secondary sludge
    b. Be unable to convert ammonia to nitrate because of the BOD load
    c. Contain less grit and inert material than a digester with only secondary sludge
    d. Produce volatile fatty acids and remove phosphorus biologically

17. Nitrification in an aerobic digester may
    a. Increase alkalinity
    b. Lower the pH
    c. Use up excess BOD
    d. Enhance settling

18. For denitrification to occur in an aerobic digester
    a. DO must be less than 0.3 mg/L
    b. Influent must be added to increase BOD
    c. All of the ammonia must be converted to nitrate
    d. pH must be below 6.5 s.u.

19. Doing this might help prevent alkalinity and pH drop in an aerobic digester
    a. Maintain DO of at least 1.5 mg/L
    b. Addition of hydrochloric or sulfuric acid
    c. Cycling between aerobic and anoxic conditions
    d. Using submerged mixers to keep solids suspended

20. Air-off cycles should be ended when
    a. Settled solids rise to the top of the digester
    b. Solids have settled to the bottom of the digester
    c. pH is near neutral and alkalinity has increased
    d. Ammonia concentrations exceed 50 mg/L as N

21. When primary sludge is added to an aerobic digester
    a. Oxygen demand increases
    b. Secondary treatment costs increase
    c. Digester size decreases
    d. Operating costs decrease

22. The primary purpose of a primary clarifier is to
    a. Remove rags, grit, and scum
    b. Decrease BOD load to downstream processes
    c. Reduce overall oxygen demand
    d. Produce volatile fatty acids

Design Parameters and Expected Performance

1. An operator may need to add more air than is needed to achieve a DO of 2 mg/L simply to keep the solids from settling to the floor of the digester.

   ☐ True
   ☐ False

2. The supernatant always collects at the very top of an aerobic digester.

   ☐ True
   ☐ False

3. Fecal coliform bacteria are typically pathogenic.

   ☐ True
   ☐ False

4. Aerobic digesters meet the definition of a PSRP when

   a. The SRT is at least 40 days at 20 °C (68 °F)
   b. The SRT is at least 60 days at 10 °C (68 °F)
   c. The SRT is at least 40 days at 15 °C (59 °F)
   d. The SRT is at least 60 days at 5 °C (86 °F)

5. The minimum VS reduction required to meet the VAR under the 503 regulations is

   a. 27%
   b. 38%
   c. 52%
   d. 61%

6. This is the maximum expected solids concentration achievable through decanting alone (no chemical addition) in an aerobic digester.

   a. 0.5%
   b. 1.2%
   c. 1.75%
   d. 3.5%

7. Which set of conditions will require the highest airflow to achieve a DO concentration of 2 mg/L in the digester?

   a. Cold water, high solids
   b. Cold water, low solids
   c. Warm water, high solids
   d. Warm water, low solids

8. An aerobic digester operates with an SRT of 20 days and achieves a VSR of 16%. It would most likely be classified as

   a. PSRP
   b. Solids holding tank
   c. Process to further reduce pathogens
   d. Continuous decanting

9. The concentration of nitrogen in aerobic digester supernatant depends on all of the following EXCEPT

   a. Percentage of primary sludge added
   b. Cycling between aerobic and anoxic conditions
   c. Percentage of volatile solids reduction
   d. Activity of nitrifying and denitrifying bacteria

Equipment

1. An advantage of operating two digesters in series instead of parallel is that it
    a. Reduces the likelihood of pathogen pass-through
    b. Improves VS reduction by 20%
    c. Reduces oxygen consumption in process
    d. Requires less mixing than parallel operation

2. This feature of aerobic digesters makes it easier to remove solids from the bottom:
    a. Integral covers
    b. Sidewater depth
    c. Sloped floor
    d. Decant hatch

3. The purpose of the aeration system is to
    a. Mix the digester contents
    b. Prevent solids from settling
    c. Provide oxygen to microorganisms
    d. All of the above

4. This type of aeration system has a greater potential for foaming than others.
    a. Fine-bubble diffused air
    b. Surface aerators
    c. Coarse-bubble diffused air
    d. Submerged mixer

5. Digested solids are typically removed from an aerobic digester at this location.
    a. Right below the water surface
    b. Next to the sludge inlet pipe
    c. Near the bottom
    d. Middle of the blanket

Hydraulic Detention Time and Solids Retention Time

1. For aerobic digesters operated without decanting, the HDT and the SRT are the same.
    ☐ True
    ☐ False

2. After a decant cycle, the water level in the digester will be lower and the HDT will be
    a. Longer
    b. Shorter
    c. Approximately the same
    d. Unchanged

3. Find the volume of a digester that is 15.2 m (50 ft) long by 9.1 m (30 ft) wide, and 3.7 m (12 ft) deep.
    a. 154 m³ (5460 cu ft)
    b. 506 m³ (18 000 cu ft)
    c. 800 m³ (28 250 cu ft)
    d. 1660 m³ (58 617 cu ft)

4. The digester can hold 1528 m³ (54 000 cu ft) of digesting solids at its highest fill level. Each day, 170 m³/d (45 000 gpd) of feed sludge is added to the digester and an equal volume of digested sludge and decant are removed. What is the HDT in days?
    a. 2.8 days
    b. 3.7 days
    c. 7.2 days
    d. 9.0 days

5. Feed sludge is added to an aerobic digester at a rate of 2.08 L/s (33 gpm). If the rate is constant over the entire day, how many cubic meters per day (gallons per day) of feed sludge will be added to the digester?

    a.   127 m³/d (33 500 gpd)

    b.   159 m³/d (42 000 gpd)

    c.   180 m³/d (47 500 gpd)

    d.   208 m³/d (55 000 gpd)

6. Find the HDT for an aerobic digester with the following characteristics. Digester is 24.4 m (80 ft) long, 15.2 m (50 ft) wide, and 5.49 m (18 ft) deep. Feed sludge is added to the digester at a rate of 47.3 m³/d (12 500 gpd). The digester is operated at its maximum fill level. Digested solids and decant are removed from the digester at the same rate that feed sludge is added so the water level does not change.

    a.   34.2 days

    b.   43.1 days

    c.   52.3 days

    d.   57.4 days

7. Supernatant is routinely removed from an aerobic digester by decanting. Which of the following statements is true?

    a.   The SRT and HDT are the same.

    b.   The HDT is longer than the SRT.

    c.   The SRT is longer than the HDT.

    d.   The HRT is shorter than the HDT.

8. One reason for high TSS concentrations in aerobic digester supernatant might be

    a.   Absence of filamentous bacteria

    b.   Nitrification during digestion

    c.   Overly long settling times

    d.   Denitrification during decanting

9. An aerobic digester contains 1634 m³ (0.4308 mil. gal) of digesting solids at 3.0% TS. How many kilograms (pounds) of solids does the digester contain?

    a.   24 510 kg (50 118 lb)

    b.   34 925 kg (76 975 lb)

    c.   43 862 kg (96 672 lb)

    d.   49 020 kg (107 786 lb)

10. An aerobic digester contains 122 580 kg (269 440 lb) of digesting solids. Each day, 2460 kg (5404 lb) of digested solids are removed from the digester. Another 25.8 kg (56.7 lb) of solids leave in the decant. What is the SRT?

    a.   24.5 days

    b.   32.1 days

    c.   49.3 days

    d.   60.7 days

11. An aerobic digester contains 114 408 kg (251 524 lb) of digesting solids. How many kilograms (pounds) can be removed from the digester each day while still meeting a minimum SRT of 40 days?

    a.   2860 kg (6288 lb)

    b.   28 600 kg (62 880 lb)

    c.   3250 kg (7163 lb)

    d.   32 500 kg (71 630 lb)

12. Given the following information, find the SRT. Digester currently holds 3814 m³ (1.01 mil. gal) of digesting solids, with a concentration of 2.8% TS. Finished biosolids are removed from the digester continuously at a rate of 0.75 L/s (12 gpm). The digester is decanted daily. Laboratory results show the decant contains 5000 mg/L of TS. On average, 67.1 m³/d (17 720 gpd) of decant are removed.

    a.   36.7 days

    b.   49.4 days

    c.   58.1 days

    d.   60.2 days

13. Denitrification in the digester caused some of the solids in the digester to rise to the surface during decanting. Using pumping records and laboratory results, the operator estimated that 293.7 m³ (77 600 gal) of decant with an average concentration of 1% TS were returned to the secondary treatment process. Before the decant started, the digester contained 17 005 m³ (4.5 mil. gal) of digesting solids at 1.8% TS. What will the SRT be if 8334 kg (18 375 lb) of digested solids are removed from the digester?

    a. 27.2 days

    b. 38.3 days

    c. 47.6 days

    d. 51.1 days

Other Process Variables

1. The SLR and SRT are related.

    ☐ True

    ☐ False

2. When calculating the SLR, the TS of the feed sludge is used.

    ☐ True

    ☐ False

3. The mass formula may be modified to calculate either kilograms (pounds) or kilograms per day (pounds per day) depending on whether basin volume or flowrate entering or leaving the basin is used.

    ☐ True

    ☐ False

4. Digester fill level has no effect on the VS loading rate.

    ☐ True

    ☐ False

5. If the pH is known, the alkalinity concentration can be calculated.

    ☐ True

    ☐ False

6. An aerobic digester is 91.4 m (300 ft) long, 30.5 m (100 ft) wide, and 6.1 m (20 ft) deep. Each day, 757.1 m³ (200 000 gpd) of WAS is added to the digester. The thickened WAS contains 4.2% TS and is 80% volatile. How many kilograms per day (pounds per day) of VS are added to the digester each day?

    a. 22 806 kg/d (50 226 lb/d)

    b. 25 439 kg/d (56 045 lb/d)

    c. 28 508 kg/d (62 832 lb/d)

    d. 31 798 kg/d (70 056 lb/d)

7. An aerobic digester is 91.4 m (300 ft) long, 30.5 m (100 ft) wide, and 6.1 m (20 ft) deep. Each day, 757.1 m³ (200 000 gpd) of WAS is added to the digester. The thickened WAS contains 4.2% TS and is 80% volatile. What is the VS loading rate to the digester?

    a. 1.5 kg/m³·d (0.09 lb/cu ft·d)

    b. 1.8 kg/m³·d (0.11 lb/cu ft·d)

    c. 3.0 kg/m³·d (0.18 lb/cu ft·d)

    d. 3.6 kg/m³·d (0.22 lb/cu ft·d)

8. Volatile solids loading rates to aerobic digesters should be held as constant as possible to

    a. Increase foaming

    b. Maximize odors

    c. Ease calculations

    d. Minimize upsets

9. An aerobic digester received 12 000 kg (5500 lb) of VS. After 40 days, only 7200 kg (3300 lb) remained. What percentage of VS were destroyed?

    a. 30%

    b. 40%

    c. 60%

    d. 80%

10. An aerobic digester is fed 15 000 kg (6800 lb) of TS that are 75% volatile. After digestion is complete, how many kilograms (pounds) of inert solids will remain?

    a. 3750 kg (1700 lb)

    b. 7500 kg (3400 lb)

    c. 11 250 kg (5100 lb)

    d. 13 000 kg (5900 lb)

11. An aerobic digester receives a mixture of primary and secondary sludge that contains 82% VS. The digested solids leaving the digester contain 73% VS. Calculate the percentage of VS reduction through the digester.

    a. 9.0%

    b. 38.2%

    c. 40.7%

    d. 59.9%

12. Digesters typically don't achieve more than approximately 50% VSR because

    a. Nonvolatile solids in the feed sludge aren't biodegradable.

    b. Some VS in the feed sludge aren't biodegradable.

    c. 50% is the limit of the VSR calculation.

    d. Most digesters don't have long enough SRTs.

13. A digester is operated at 22 °C (71.6 °F) and an SRT of 42 days. How many degree-days is this?

    a. 528

    b. 924

    c. 1800

    d. 3007

14. Aerobic digesters perform best at a pH closest to

    a. 4

    b. 7

    c. 10

    d. 12

15. An aerobic digester is considered to be sour when

    a. Alkalinity is greater than 100 mg/L

    b. SLR is greater than 0.3 lb BOD/cu ft·d

    c. pH falls below 6.5 s.u.

    d. CO_2 concentration exceeds 400 ppm

16. One method for preventing further pH drop in an aerobic digester is to

    a. Add hydrochloric or citric acid

    b. Increase solids retention time

    c. Ensure hatches on covered tanks are closed

    d. Lower DO concentration to inhibit nitrification

17. Cycling an aerobic digester between aerobic and anoxic conditions

    a. Lowers nitrate concentrations

    b. Increases overall energy costs

    c. Decreases supernatant alkalinity

    d. Accentuates VSR

18. A side effect of operating aerobic digesters at DO concentrations below 1.0 mg/L is
    a. Increased ammonia removal
    b. Decreased VS destruction
    c. Increased nitrate production
    d. Improved settling during decant

19. Decanting can increase the TS concentration in a digester to approximately _____ without the addition of chemicals.
    a. 0.5%
    b. 2.0%
    c. 3.5%
    d. 5.0%

20. A digester contains 17 005 m³ (600 000 cu ft) of digesting solids with a TS concentration of 1.5%. The digester contents are allowed to settle and 7895 m³ (278 571 cu ft) of supernatant is removed from the digester. After aeration and mixing are resumed, what will the new TS concentration be in the digester?
    a. 0.8%
    b. 1.9%
    c. 2.6%
    d. 2.8%

21. The solids concentration in an aerobic digester is 3.5%. The operator has turned the air off for several hours, but can't get a supernatant layer to form. What is the most likely reason?
    a. Maximum concentration is already achieved
    b. Solids particles are too large to compact
    c. Low pH is preventing coagulation from occurring
    d. Water temperature is hindering settling

22. Typically, the air and mixing only need to be turned off in a particular digester for a few hours before several feet of supernatant accumulate. Today, the air has been off for 12 hours, but the solids are not settling. The solids concentration in the digester is only 1.5% TS. What is the most likely cause?
    a. Water temperature is too warm
    b. Filamentous bacteria are present
    c. Shorter SRT and larger particles
    d. Operator forgot to add lime

Process Control

1. Most WRRFs return digester decant continuously.
    ☐ True
    ☐ False

2. Slug loading an aerobic digester with VS
    a. May cause foaming
    b. Increases overall SRT
    c. Reduces decant frequency
    d. Causes rag accumulation

3. Digester decant is typically returned to the liquid stream
    a. Upstream of the headworks
    b. After influent flow monitoring
    c. Directly to the aeration basin
    d. At the primary splitter box

4. Decant containing 40 mg/L of TP is returned to the primary clarifier effluent splitter box. The primary clarifier effluent contains 6 mg/L of TP. Find the TP concentration in the combined flows if the influent flow is 45 420 m³/d (12 mgd) and the decant return flow is 9538 m³/d (2.52 mgd).

    a. 7.5 mg/L
    b. 9.0 mg/L
    c. 10.2 mg/L
    d. 11.9 mg/L

5. An anaerobic digester is fed with a mixture of primary and secondary sludge. The primary sludge contains 6% TS. The secondary sludge contains 0.8% solids. If 113.55 m³/d (30 000 gpd) of primary sludge and 189.25 m³/d (50 000 gpd) of secondary sludge are blended together, what will the feed sludge concentration be to the digester?

    a. 1.83%
    b. 2.75%
    c. 4.05%
    d. 5.21%

6. Returning digester supernatant with a high TSS concentration

    a. Decreases overall operating costs
    b. Affects clarifier's surface overflow rate
    c. May increase effluent turbidity
    d. Affects BOD removal rates

7. The best time to return supernatant would be

    a. During periods of low flow
    b. When the facility is fully staffed
    c. Early afternoon or evening
    d. Whenever the digester is operating

Operation

1. When starting up a new digester or one that has been out of service, the digester should be filled completely with feed sludge before starting the aeration system.

    ☐ True
    ☐ False

2. Starting up a digester with primary sludge takes longer than starting up a digester with secondary sludge.

    ☐ True
    ☐ False

3. The surface of an aerobic digester has undisturbed foam in one quadrant. The rest of the digester surface is nearly foam free, with air bubbles breaking the surface in a regular pattern. The operator concludes the following:

    a. The quadrant with foam has a broken header.
    b. The quadrant with foam is missing a diffuser.
    c. The quadrant with foam has clogged diffusers.
    d. The quadrant with foam has good mixing.

4. A well-operated aerobic digester should smell

    a. Pleasant and musty
    b. Sharp and sour
    c. Like rotten eggs
    d. Fruity and hoppy

5. To ensure adequate VS destruction, the DO concentration in a digester should be
   a. Less than 0.5 mg/L
   b. At least 1.0 mg/L
   c. Equal to 2.0 mg/L
   d. Greater than 2.0 mg/L

6. Solids settling and decant phases should be limited to 3 to 4 hours to
   a. Minimize floating solids
   b. Maximize denitrification
   c. Prevent diffuser clogging
   d. Reduce liquid stream effects

7. Pumps and sludge lines should be emptied of sludge by running clean water through them before long-term storage because
   a. Trapped sludge may produce gases and pressure
   b. Dried sludge can damage equipment
   c. Trapped sludge may dry, forming a blockage
   d. It is easier to remove fresh sludge than older sludge

Monitoring, Maintenance, Troubleshooting, and Safety

1. When conducting a SOUR test on an aerobic digester, the digester TS concentration should be used.
   ☐ True
   ☐ False

2. High turbidity in digester samples can interfere with colorimeter-based laboratory tests.
   ☐ True
   ☐ False

3. When entering a digester to perform cleaning and maintenance, operators should first determine if the digester is a confined space.
   ☐ True
   ☐ False

4. Eating, drinking, and smoking are all permissible near biological treatment processes.
   ☐ True
   ☐ False

5. Supernatant is sampled and analyzed to
   a. Maintain compliance with biosolids permit
   b. Predict impacts to liquid stream processes
   c. Determine VS destruction
   d. Estimate degree of solids stabilization

6. Samples taken at this location are necessary for calculating the VS loading rate.
   a. Feed sludge
   b. Process
   c. Digested solids
   d. Supernatant

7. The SOUR result for an aerobic digester is 0.8 mg/L DO/g TS/h. This indicates
   a. Excess BOD remains
   b. Presence of ammonia
   c. Solids are stabilized
   d. Aeration is too low

8. How often should the aeration equipment in an aerobic digester be inspected?
    a. Daily
    b. Weekly
    c. Monthly
    d. Yearly

9. Increased blower discharge pressure may indicate
    a. Fouled diffusers
    b. Broken air header
    c. Missing diffusers
    d. Lower water level

10. Decanting and solids withdrawal times should be kept as short as possible to
    a. Reduce operator workload
    b. Maximize decant volume
    c. Prevent diffuser fouling
    d. Ensure maximum solids concentrations

11. Oxygen transfer efficiency and resulting DO concentrations may be reduced by
    a. Water temperatures below 15 °C (59 °F)
    b. Total solids concentrations >3.5%
    c. Specific oxygen uptake rate
    d. Nitrate and nitrite accumulation

12. Low DO concentrations and odors in an aerobic digester may be caused by
    a. Extended SRTs
    b. High VS loading rate
    c. Increased oxygen transfer efficiency
    d. Filamentous bacteria and foaming

CHAPTER 4
Anaerobic Digestion

Theory of Operation

1. Anaerobic digestion is typically used at larger WRRFs that have permitted capacities greater than approximately 19 000 m³/d (5 mgd).
   - ☐ True
   - ☐ False

2. Biogas produced in anaerobic digesters may be used to heat the digester or to produce electricity.
   - ☐ True
   - ☐ False

3. Anaerobic digesters are typically fed with secondary sludge only.
   - ☐ True
   - ☐ False

4. All anaerobic digesters are capable of achieving 50% VSR.
   - ☐ True
   - ☐ False

5. Methanogens always use acetic acid to produce methane gas.
   - ☐ True
   - ☐ False

6. This valuable nutrient can be recovered from anaerobic digesters.
   a. Potassium
   b. Calcium
   c. Phosphorus
   d. Nitrate

7. Digested solids are considered to be stabilized when they
   a. Meet requirements in the 503 regulations
   b. Anaerobically digest for at least 10 days
   c. Contain less than 70% VS
   d. No longer contain any pathogens

8. Anaerobic digesters are covered to prevent
   a. Odor generation
   b. Air from entering
   c. Windblown debris
   d. Sludge overflows

9. One goal of anaerobic digestion is to
   a. Increase vector attraction
   b. Minimize the mass of inert solids
   c. Produce carbon dioxide gas
   d. Reduce the mass of VS

10. This term is used to describe the material entering a digester.
    a. Feed sludge
    b. Solids
    c. Biosolids
    d. Primary

11. Inert material entering an anaerobic digester
    a. Breaks down during hydrolysis
    b. Serves as a food source for methanogens
    c. Passes through the digester unchanged
    d. Collects in the digester supernatant

12. The rate-limiting step of anaerobic digestion is
    a. Hydrolysis
    b. Acidogenesis
    c. Acetogenesis
    d. Methanogenesis

13. The acidogenic bacteria convert
    a. Volatile fatty acids into acetic acid
    b. Soluble organic compounds into VFAs
    c. Starches into simple sugars
    d. Organic matter into methane

14. Ammonia generated through the breakdown of VS
    a. Lowers the pH of the digester supernatant
    b. Reacts with calcium to form a precipitate
    c. Combines with carbon dioxide to raise alkalinity
    d. Assists in formation of vivianite crystals

15. Magnesium ammonium phosphate is more commonly known as
    a. Struvite
    b. Vivianite
    c. Chert
    d. Blue stone

16. In anaerobic digesters, this group of bacteria is most likely to be harmed by toxic compounds entering the digester.
    a. Acidogens
    b. Acetogens
    c. Methanogens
    d. Sulfate reducers

Design Parameters and Expected Performance

1. Conventional mesophilic digesters are operated at temperatures near
    a. 25 °C (77 °F)
    b. 35 °C (95 °F)
    c. 45 °C (113 °F)
    d. 55 °C (131 °F)

2. This design and operating parameter determines whether a digester is considered low or high rate.
    a. Operating temperature
    b. Type of feed sludge
    c. Solids retention time
    d. Volatile solids loading rate

3. The SRT for a conventional, mesophilic digester is typically
    a. 10 days
    b. 20 days
    c. 30 days
    d. 40 days

4. One advantage of thermophilic digestion over mesophilic digestion is
    a. Greater pathogen destruction
    b. Lower VS loss
    c. Larger overall tank size
    d. Simpler overall operation

5. Most anaerobic digesters achieve _____ VSR.
    a. 10%–20%
    b. 25%–35%
    c. 45%–55%
    d. 65%–75%

6. Anaerobic digesters are capable of producing
    a. Class A biosolids
    b. Class B biosolids
    c. Class C biosolids
    d. Class D biosolids

7. Biogas produced by anaerobic digestion typically contains _____ % methane.
    a. 35
    b. 50
    c. 65
    d. 80

8. The energy value of digester biogas is lower than the energy value of natural gas because biogas contains a high percentage of
    a. Siloxanes
    b. Hydrogen sulfide
    c. Struvite
    d. Carbon dioxide

9. When siloxanes are burned, they form this abrasive compound, also known as beach sand.
    a. Silicon dioxide
    b. Struvite
    c. Vivianite
    d. Sodium silicate

10. Ammonia concentrations are higher in anaerobic digester supernatant than in aerobic digester supernatant because
    a. Anaerobic digesters achieve more VS destruction
    b. Aerobic digesters support the growth of nitrifying bacteria
    c. Ammonia combines with carbon dioxide in anaerobic digesters
    d. Struvite formation only occurs in anaerobic digesters

Equipment

1. Digester tank shape has the largest direct effect on the _____ of the digester.
    a. Mixing characteristics
    b. Microbiological makeup
    c. Heat retention
    d. Solids retention time

2. _____ digester tanks become more favorable for digester projects at facilities that have extra land area available and limited construction budgets.
   a. Egg-shaped
   b. Pancake
   c. Silo
   d. Square

3. The exterior membrane of a dual membrane cover will rise and fall depending on how much biogas is being stored within the cover.
   ☐ True
   ☐ False

4. An ideal digester maintains plug flow through the tank.
   ☐ True
   ☐ False

5. Gas mixing systems use compressed _____ injected to the anaerobic digester liquid to create mixing currents within the tank.
   a. Nitrogen gas
   b. Biogas
   c. Air
   d. Chlorine gas

6. The greatest portion of a digester's heat demand typically comes from _____.
   a. Heat loss through the cover
   b. Heat loss through the walls
   c. Heating the incoming sludge
   d. Heating the biogas

7. Most heating systems for mesophilic digesters are designed to run intermittently, turning on when the digester temperature hits 30 °C (85 °F) and the turning off when the temperature reaches 38 °C (100 °F).
   ☐ True
   ☐ False

8. A benefit of using an ESD is
   a. Low construction cost
   b. Internal gas storage capability
   c. Good mixing characteristics
   d. Variable water level operation

9. Air is excluded from a floating cover digester by
   a. An accordion-type rubber seal
   b. A water seal at the bottom of the skirt
   c. A metal telescoping seal
   d. A membrane seal at the top of the wall

10. Digester covers are protected from high pressure and vacuums by
    a. A fan that continuously circulates air
    b. A relief valve
    c. A programmable logic controller based pressure control system
    d. A grated cover vent

11. Which of the following is the most important purpose of an anaerobic digester cover?
    a. Keeping debris out of the tank
    b. Keeping UV light out of the tank
    c. Keeping oxygen out of the tank
    d. Keeping chemicals out of the tank

12. In digesters with floating covers, corbels serve which of the following purposes?
    a. Keep the liquid in the tank
    b. Prevent the cover from damaging internal equipment
    c. Create a bubble to contain biogas
    d. Diffuse chemicals to control pH

13. _____ studies are conducted by placing a precise amount of inert, detectable material in the digester to determine mixing system performance.
    a. Thermal
    b. Current
    c. Mixing
    d. Tracer

14. Pumped mixing systems are advantageous because the mechanical equipment is on the outside of the digester and, therefore, more easily accessible.
    ☐ True
    ☐ False

15. The goal of sludge screening is to
    a. Increase the rate of digestion
    b. Decrease the energy needed for pumping
    c. Decrease the debris that collects in the digester
    d. Increase the VS in the feed sludge

16. The goal of adding sludge pretreatment processes such as thermal hydrolysis before anaerobic digestion is to
    a. Increase the rate of digestion
    b. Decrease the energy needed for pumping
    c. Decrease the debris that collects in the digester
    d. Increase the inert solids in the feed sludge

17. Feed sludge should be added
    a. Immediately before the heat exchanger
    b. Immediately after the heat exchanger
    c. Adjacent to the discharge point
    d. At the water surface

18. An observation that indicates that proper mixing is occurring in the digester is
    a. High gas production
    b. Scum formation at the water surface
    c. pH trending downward
    d. Grit accumulation in the bottom

19. The purpose of draft tubes on mechanical mixers is to
    a. Reduce energy use of the mixer
    b. Create velocity currents close to the bottom of the tank
    c. Create a mixing vortex
    d. Provide a solid support for the mixer

20. To support a stable digestion process, feed sludge should be
    a. Refrigerated to preserve the organic solids
    b. Aerated to increase the dissolved oxygen concentration
    c. Evaporated to reduce water content
    d. Fed slowly to avoid overloading the methanogens

Process Variables and Process Control

1. Anaerobic digestion generates alkalinity.
   - ☐ True
   - ☐ False

2. As alkalinity increases, digester pH decreases.
   - ☐ True
   - ☐ False

3. Volatile fatty acids are toxic to methanogens at low concentrations.
   - ☐ True
   - ☐ False

4. Increasing the VS loading rate to an anaerobic digester by more than 10% per day can trigger a foaming event.
   - ☐ True
   - ☐ False

5. Which of the following are included in the working volume of an anaerobic digester?
   - a. Settled inert solids
   - b. Scum and debris
   - c. Digesting solids
   - d. Entire digester volume

6. Solids retention times for mesophilic anaerobic digesters are typically approximately
   - a. 10 days
   - b. 20 days
   - c. 30 days
   - d. 40 days

7. Solids leave the digester through
   - a. Sludge withdrawal and decanting
   - b. Wasting and sludge withdrawal
   - c. Overflow piping and decanting
   - d. Sludge withdrawal only

8. Washout of methanogens occurs when
   - a. SRT is too long
   - b. HDT is too short
   - c. SRT is too short
   - d. HDT is too long

9. A primary anaerobic digester holds 100 025 kg (220 370 lb) of digesting sludge. Each day, 3925 kg (8650 lb) are removed. The digester is not decanted. What is the SRT?
   - a. 11.6 days
   - b. 15.3 days
   - c. 25.5 days
   - d. 56.1 days

10. The feed sludge contains 2.3% TS. What is this in milligrams per liter (mg/L)?
    - a. 230
    - b. 2300
    - c. 23 000
    - d. 230 000

11. A primary anaerobic digester has a maximum volume of 5557 m³ (196 250 cu ft). It is fed continuously at 3.15 L/s (50 gpm). The feed sludge has a TS concentration of 4.3% and contains 82% VS. Find the VS loading rate.

    a. 1.33 kg/m³·d (0.08 lb/cu ft·d)

    b. 1.68 kg/m³·d (0.10 lb/cu ft·d)

    c. 2.05 kg/m³·d (0.13 lb/cu ft·d)

    d. 2.86 kg/m³·d (0.18 lb/cu ft·d)

12. The temperature of a mesophilic anaerobic digester should not change by more than _____ each day.

    a. 0.3 °C (0.5 °F)

    b. 0.6 °C (1.0 °F)

    c. 1.0 °C (1.8 °F)

    d. 2.0 °C (3.6 °F)

13. This form of nitrogen can become toxic to methanogens, especially at pH values above 8.0 s.u.

    a. Ammonium bicarbonate

    b. Ionized ammonia (NH_4^+)

    c. Ammonium hydroxide

    d. Free ammonia (NH_3)

14. A sludge sample is collected from a primary digester. If it isn't analyzed within 10 minutes

    a. pH may decrease

    b. Inerts will increase

    c. pH may increase

    d. Inerts will decrease

15. The volatile acids concentration in an anaerobic digester is 275 mg/L. The alkalinity concentration is 1750 mg/L as $CaCO_3$. What is the VA/ALK?

    a. 0.16

    b. 0.32

    c. 0.75

    d. 2.75

16. One indicator that a toxic compound has entered an anaerobic digester is

    a. Increased methane gas production

    b. Decreased digester operating temperature

    c. Increased volatile acid concentration

    d. Decreased SLR

17. The VS loading rate to an anaerobic digester should be

    a. Allowed to fluctuate by 20%

    b. Kept as constant as possible

    c. Flow-paced to the influent

    d. Adjusted based on the VA/ALK

18. The VS loading rate to an anaerobic digester increases by 30% in a single day. Which of the following is most likely to occur?

    a. Increase in methane gas production

    b. pH increases to over 8.0 s.u.

    c. Death of acetogenic bacteria

    d. Accumulation of volatile acids

19. Which digester feeding schedule is the least likely to result in a process upset?

    a. Feed continuously

    b. Feed 15 minutes each hour

    c. Feed 3 times per day

    d. Feed once per day

20. Withdrawing sludge and supernatant from a fixed cover digester faster than feed sludge is added may result in
    a. Pressure buildup in the digester
    b. Explosive mixture of air and biogas
    c. Violation of the Clean Air Permit
    d. Excessively long SRT

21. Sludge is typically prethickened before adding it to an anaerobic digester to
    a. Reduce the DT
    b. Increase heating requirements
    c. Maximize solids handling
    d. Prevent alkalinity loss

22. The most common cause of digester foaming is
    a. Organic overloading
    b. Surfactants
    c. Filamentous bacteria
    d. VFA accumulation

23. A rapid rise event in an anaerobic digester
    a. Caused by over or under mixing
    b. Creates a surface foam layer
    c. Results from chronic underfeeding
    d. Separates biogas from the solids

Operation

1. When starting up a new anaerobic digester or one that has been out of service, the digester should be filled completely with raw feed sludge before checking the readiness status of the equipment.
    ☐ True
    ☐ False

2. Solids should be removed from the digester immediately before feeding the digester to prevent _____.
    a. Gas development
    b. Short-circuiting
    c. Volatile solids reduction
    d. Increased alkalinity

3. Unless you decant the digester, TS concentration in the sludge discharge will likely be less than the feed solids because VS are destroyed during the anaerobic digestion process.
    ☐ True
    ☐ False

4. Anaerobic digesters perform well when they are batch fed consistent volumes of sludge once every 48 hours.
    ☐ True
    ☐ False

5. Decanting is performed in order to _____ and increase the SRT in the digester.
    a. Increase the VS concentration
    b. Decrease the VS concentration
    c. Increase the TS concentration
    d. Decrease the TS concentration

6. _____ should slowly be reduced for weeks ahead of time when planning to take an anaerobic digester offline.
   a. Volatile solids loading
   b. Mixing intensity
   c. Tank water level
   d. Digester temperature

7. Continuous digester mixing is important to _____.
   a. Maintain the liquid level in the tank
   b. Prevent scum from floating to the top of the tank
   c. Prevent low dissolved oxygen pockets in the sludge
   d. Avoid formation of struvite

8. The digester heating system should stay in operation during a decant cycle to prevent the temperature from dropping more than 0.6 °C (1 °F).
   ☐ True
   ☐ False

9. The important parameter to monitor to ensure proper digester feeding is _____.
   a. Temperature
   b. Feed volume
   c. pH
   d. Feed VS load

10. _____ is added to enclosed spaces where biogas could accumulate to reduce the risk of a fire or explosion.
    a. Air
    b. Oxygen
    c. Nitrogen
    d. Hydrogen

11. A long-term approach to managing foam in the digester would be to _____.
    a. Identify an effective chemical defoamer
    b. Install additional spray headers
    c. Identify the root cause and correct it
    d. Install additional emergency relief vents

12. Mixing systems should be designed to break up scum mats at the surface of the digester and integrate them back into the rest of the liquid in the tank.
    ☐ True
    ☐ False

Data Collection, Sampling, and Analysis and Maintenance

1. Which two samples must be collected to calculate the VSR through an anaerobic digester?
   a. Feed sludge and digested sludge
   b. Digested sludge and supernatant
   c. Primary and secondary digester
   d. Supernatant and feed sludge

2. Before a sample is collected, _____.
   a. Sample lines should be flushed
   b. Crack sampling port to 50% open
   c. Rinse container with hot water
   d. Turn off mixers and pumps

3. Feed sludge VS should be measured
   a. In the primary sludge only
   b. In the secondary sludge only
   c. At the entry point to each digester
   d. After blending sludge sources

4. Laboratory results should be
   a. Locked in the supervisor's office
   b. Charted to identify trends
   c. Recorded in multiple locations
   d. Reviewed only during upsets

5. The goal of a formalized maintenance program is to _____.
   a. Eliminate the need to replace equipment
   b. Document maintenance costs for warranty claims
   c. Improve the reliability of critical equipment
   d. Assign responsibility if equipment fails

6. _____ can increase HRT and improve VSR.
   a. Adding surface sprayers
   b. Cleaning the digester
   c. Removing fouling from a heat exchanger
   d. Removing struvite from a pump

7. A gradual decline in the heat transfer rate in a digester heating system is likely a sign of _____.
   a. A boiler failure
   b. A clogged hot water pump
   c. A fouled heat exchanger
   d. A damaged temperature transmitter

Safety Considerations

1. It is possible for small amounts of biogas to be released from pressure-relief valves during normal operation of an anaerobic digester.
   ☐ True
   ☐ False

2. Inhalation exposure to which component of biogas poses the most serious health risks?
   a. Methane
   b. Hydrogen sulfide
   c. Carbon dioxide
   d. Oxygen

3. An area where there is less than 21% oxygen is considered an oxygen-deficient atmosphere.
   ☐ True
   ☐ False

4. Flame arresters protect against _____.
   a. Backflash fires
   b. Ignition
   c. Combustion
   d. Flashover fires

5. The three things that must be present at the same time for a fire to occur include oxygen, heat, and _____.

    a. Wood

    b. Open flame

    c. Fuel

    d. Carbon dioxide

6. The presence of methane gas can be detected by

    a. Observing its characteristic green color

    b. A methane sensor mounted at the lowest point in the area

    c. Smelling its pungent "rotten egg" odor

    d. An LEL sensor set up for methane

CHAPTER 5
Dewatering

Purpose and Function of Dewatering

1. Dewatering reduces the total mass of sludge.
   - ☐ True
   - ☐ False

2. The belt filter press is the best equipment option for dewatering chemical sludges.
   - ☐ True
   - ☐ False

3. Drying beds are the oldest means of sludge dewatering.
   - ☐ True
   - ☐ False

4. Heat drying and pelletization of sludge is used to create a beneficial reuse fertilizer that can be commercially sold.
   - ☐ True
   - ☐ False

5. Chemical conditioning is typically not required in sludge dewatering.
   - ☐ True
   - ☐ False

Centrifuges for Dewatering

1. *Solids capture rate* refers to the amount of solids that are able to be retained in the biosolids versus the solids that leave the machine in the centrate.
   - ☐ True
   - ☐ False

2. In a dewatering centrifuge, solids separation only takes place in the conical end.
   - ☐ True
   - ☐ False

3. Which of the following can be done to increase the capacity of the centrifuge?
   a. Run the centrifuge for a longer period of time
   b. Decrease the pond depth
   c. Decrease the feed rate
   d. Run the centrifuge for a shorter period of time

4. Differential speed of the bowl to the scroll typically ranges from _____.
   a. 1–20 rpm
   b. 20–40 rpm
   c. 40–60 rpm
   d. 60–80 rpm

5. Dewatering centrifuge bowl speeds range from _____.
   a. 100–300 rpm
   b. 300–900 rpm
   c. 600–1500 rpm
   d. 1000–3000 rpm

6. Dewatering centrifuges typically produce TS in which of the following ranges?
   a. 2%–15%
   b. 5%–35%
   c. 15%–75%
   d. 25%–85%

7. Centrifuges are only effective in dewatering waste activated sludge.
   ☐ True
   ☐ False

8. The higher the percent capture, the lower the percent solids that will be returned to the head of the WRRF.
   ☐ True
   ☐ False

9. Increasing the feed rate will ensure a higher percent solids in the cake produced in a dewatering centrifuge.
   ☐ True
   ☐ False

Belt Filter Presses

1. The low-pressure zone is also known as the wedge zone.
   ☐ True
   ☐ False

2. Chemical sludges dewater easily on BFPs.
   ☐ True
   ☐ False

3. Dilute feed solids require less detention time on the gravity section of the belt.
   ☐ True
   ☐ False

4. Typical capture rates for BFPs are in which of the following ranges?
   a. 62%–68%
   b. 72%–78%
   c. 82%–88%
   d. 92%–98%

5. The gravity drainage section of the BFP operates identically to the gravity belt thickener.
   ☐ True
   ☐ False

6. Chicanes are responsible for moving the sludge into rows on the gravity section of the BFP to allow for water to drain through the belt.
   ☐ True
   ☐ False

7. When the solids particles are forced together, changing their shape, and water is released from between the particles, it is known as _____.
   a. Cutting
   b. Squeezing
   c. Shearing
   d. Drying

8. The water that fills the spaces between the solids particles is known as _____ water.
   a. Interspacial
   b. Interstitial
   c. Intergalactical
   d. Interspectral

9. Belts of a BFP are most commonly made of _____.
   a. Aluminum
   b. Cotton
   c. Polyester
   d. Silk

10. The belt tension on a BFP is typically adjusted automatically and can be monitored with a pressure gauge.
    ☐ True
    ☐ False

11. Belt press wash water systems should operate at no less than _____.
    a. 590 kPa (85 psi)
    b. 690 kPa (100 psi)
    c. 790 kPa (115 psi)
    d. 890 kPa (130 psi)

12. Which of the following is the most important daily maintenance activity for belt press operators to do?
    a. Realign the doctor blade
    b. Tighten the belt seam
    c. Wash down the BFP
    d. Adjust the belt alignment

13. What is a potential cause of solids loss through the lateral sides of the belt in a belt filter press high pressure zone?
    a. Polymer dose is too high
    b. Wash water pressure is too high
    c. Pressure on belt is too high
    d. Pressure in gravity zone is too high

14. What is a possible cause of cake sticking to the belt of a BFP?
    a. Use of new doctor blade
    b. Polymer overdose
    c. Solids loading rate too low
    d. Belt speed to low

15. Dilute sulfuric acid solution should be used to clean the belts of a belt filter press properly.
    ☐ True
    ☐ False

16. In general, with BFP operation, thicker feed solids dewater better than thinner feed material.
    ☐ True
    ☐ False

17. Belt filter presses are the best dewatering devices for industrial solids.
    ☐ True
    ☐ False

Rotary Presses

1. Rotary presses use _____ to dewater municipal wastewater solids.
    a. Gravity and pressure
    b. Gravity and centrifugal force
    c. Pressure and centrifugal force
    d. Gravity and air flotation

2. Polymer and/or conditioning chemicals are typically not needed to achieve proper performance of a rotary press.
    ☐ True
    ☐ False

3. Characteristics of a rotary press include the following:
    a. Small footprint, no odor containment, high speeds
    b. Low speeds, larger footprint, good odor containment
    c. Low energy use, good odor containment, small footprint
    d. Larger footprint, high energy use, no odor containment

4. The three different zones within a rotary press are as follows:
    a. Flocculation, restriction, and squeezing
    b. Filtration, pressing, and restriction
    c. Flotation, pressing, and flocculation
    d. Classification, flocculation, and desiccation

5. Rotary presses are made up of modular units called channels that are connected and rotated by a single, common shaft.
    ☐ True
    ☐ False

6. It is recommended to regularly perform a _____ to ensure that you are using the correct polymer type and dose.
    a. Rise rate test
    b. Capture rate test
    c. Jar test
    d. Filtration paper test

7. The _____ mode is used to perform initial startup and system adjustments on a rotary press.
    a. Production
    b. Dewatering
    c. Recirculation/bypass
    d. Pilot testing/trial and error

8. When adjusting a rotary press, increasing the solids feed pressure will likely cause _____.
    a. A decrease in the cake production
    b. An improvement in the filtrate quality
    c. An increase in the cake solids content
    d. An increase in the solids throughput

9. Monitoring the _____ of a rotary press will help to determine the effect the recycle stream will have on the main process.
    a. Feed solids flow and solids content
    b. Filtrate flow, CBOD, and nutrients concentrations
    c. Cake quantity and VS content
    d. Centrate temperature, pH, and alkalinity

10. Adjusting the _____ helps to form the initial cake plug in the rotary press.
    a. Discharge restriction pressure
    b. Roller tension on the belts
    c. Differential speed of the scroll
    d. Rotational speed of the drum

11. Low solids content in the dewatered cake being discharged from a rotary press is most likely caused by
    a. Filtration wheel speed set too low
    b. Feed solids flow is too low
    c. Discharge pressure is too low
    d. Inlet pressure is too low

12. The _____ in a rotary press creates frictional pressure that squeezes additional water from the cake and moves it out of the press.
    a. Discharge restriction pressure
    b. Feed solids pressure
    c. High-pressure wash water
    d. Rotation of the filtration wheel

13. A typical maintenance task for a rotary press is
    a. Replace a worn filter belt
    b. Inspect and clean the washwater nozzles
    c. Refurbish a worn scroll
    d. Grease the drum bearings

Screw Presses

1. Inclined screw presses are typically designed at which of the following angles?
    a. 0 to 10°
    b. 10 to 20°
    c. 20 to 30°
    d. 30 to 40°

2. One of the disadvantages of operating a screw press is the slow rotational speed.
    ☐ True
    ☐ False

3. Screw press dewatered sludge concentrations range from 40% to 95% TS.
    ☐ True
    ☐ False

4. The hydraulic and solids loading rates for inclined screw presses are typically slightly lower than those for horizontal screw presses.
    ☐ True
    ☐ False

5. Which of the following is a benefit of the inclined screw press?
    a. Higher hydraulic loading capacity
    b. Higher solids loading capacity
    c. Higher cake solids content
    d. Higher rotational speed

6. Decreasing the auger or screw speed in a screw press increases the solids content of the discharge cake.
    ☐ True
    ☐ False

7. In general, screw presses rotate at higher speeds than centrifuges.
    ☐ True
    ☐ False

8. Increasing the screw rotation speed will do which of the following?
    a. Increase the capacity of the screw press and decrease the dewatered cake solids content
    b. Decrease the capacity of the screw press and increase the dewatered cake solids content
    c. Increase the capacity of the screw press and increase the dewatered cake solids content
    d. Decrease the capacity of the screw press and decrease the dewatered cake solids content

9. Typically, screw presses dewater first by gravity and then by pressure.
   ☐ True
   ☐ False

10. The typical capture rate of a screw press ranges from which of the following?
    a. 65%–75%
    b. 75%–85%
    c. 85%–95%
    d. Always higher than 95%

11. Screw presses are a safety concern because of their high rotational speed.
    ☐ True
    ☐ False

12. In screw press operations, increasing the screw rotation speed increases production capacity, but decreases cake solids concentration.
    ☐ True
    ☐ False

Drying Beds

1. What is the primary source of odors in drying bed applications?
   a. Inadequate polymer use
   b. Inadequate digestion
   c. Inadequate applied solids
   d. Inadequate laborers

2. In most drying bed applications, sludge should be applied at which of the following depths?
   a. 50 mm (2 in.)
   b. 100 mm (4 in.)
   c. 150 mm (6 in.)
   d. 200 mm (8 in.)

3. _____ drying beds are the simplest drying bed applications.
   a. Reed
   b. Sand
   c. Paved
   d. Vacuum-assisted

4. It is common to use drying beds to dry undigested sludge.
   ☐ True
   ☐ False

5. Drying beds work best in hot, arid climates.
   ☐ True
   ☐ False

6. Drying beds typically achieve total dry solids concentrations of at least _____?
   a. 10%
   b. 20%
   c. 30%
   d. 40%

7. Sand should never have to be replaced in a drying bed if operated properly.
   ☐ True
   ☐ False

8. Which of the following can help prevent the effects of rainfall on drying bed operations?
    a. Increased polymer usage
    b. Increased humidity
    c. Use of covers
    d. Use of front-end loaders

9. It is unnecessary to clean drying beds between applications.
    ☐ True
    ☐ False

Other Dewatering Processes

1. In pressure filter presses, the pressure that dewaters the sludge is created by
    a. Centrifugal force from a rotating scroll
    b. Mechanical squeezing between belts on rollers
    c. Pumping sludge between filter plates
    d. Slowly pushing sludge against a discharge restriction

2. Dewatering sludge in a BFP is a _____ process.
    a. Plug flow
    b. Continuous flow
    c. Automated
    d. Batch

3. _____ are wrapped around plates to trap the cake and allow filtrate to drain through the plates in a pressure filter press.
    a. Filter cloths
    b. Filter belts
    c. Wedge wire screens
    d. Membranes

Sludge Cake Conveyors

1. Belt conveyors are best used to transport material that is at least _____.
    a. 1% TS
    b. 5% TS
    c. 10% TS
    d. 15% TS

2. Screw conveyors can be designed to either push or pull dewatered cake to the discharge location.
    ☐ True
    ☐ False

3. Shafted screw conveyors are typically designed in shorter lengths to avoid or minimize using intermediate bearings because they _____.
    a. Require excessive energy to operate
    b. Create interferences for material to hang up on
    c. Require excessive structural design
    d. Excessively increase the capital cost of the system

4. Shaftless screw conveyor troughs are lined with _____.
    a. Polyester
    b. Cotton
    c. Teflon
    d. Aluminum

Dewatered Sludge Storage and Hauling

1. How many days of biosolids storage are recommended?
    a. 30–60 days
    b. 60–120 days
    c. 90–150 days
    d. 120–180 days

2. Dewatered biosolids are most often stored in tanks for long-term storage.
    ☐ True
    ☐ False

Odor Control and Safety

1. Ozone is a type of oxidant used in biosolids odor control.
    ☐ True
    ☐ False

2. Safety interlocks are not an important part of dewatering system safety considerations.
    ☐ True
    ☐ False

3. Adequate ventilation is an important safety consideration in all dewatering applications.
    ☐ True
    ☐ False

CHAPTER 6
Electrical Fundamentals and Motors

What Is Electricity?

1. Atoms are normally
   a. Positively charged
   b. Electrically neutral
   c. Negatively charged

2. The electrons in the outermost shell of an atom are the
   a. Valence electrons
   b. Lost electrons
   c. Binding electrons
   d. Covalent electrons

3. When copper is pure or nearly pure, the electrons in the outermost shell may
   a. Randomly jump between orbitals
   b. Form covalent bonds with other elements
   c. Move from one copper atom to another
   d. Generate electricity through random movement

4. The attraction between protons and electrons is related to how far apart they are.
   ☐ True
   ☐ False

5. The north and south ends of a magnet or magnetic field are called
   a. Ions
   b. Valences
   c. Attractors
   d. Poles

6. One example of a good insulator might be
   a. Copper
   b. Rubber
   c. Aluminum
   d. Iron

7. One example of a good conductor might be
   a. Styrofoam
   b. Rubber
   c. Glass
   d. Silver

8. Electricity requires the coordinated _____ of electrons.
   a. Charging
   b. Movement
   c. Potential
   d. Difference

9. Electricity may be shown flowing from the positive end of the battery (conventional theory) or from the negative end of the battery (electron theory).
    ☐ True
    ☐ False

10. Placing a magnet near a coil of copper wire generates electricity.
    ☐ True
    ☐ False

11. Current flowing through a wire generates a magnetic field.
    ☐ True
    ☐ False

12. One method for increasing the strength of a magnetic field generated by a coil of wire is to
    a. Decrease the number of coils
    b. Insert a glass rod into the coil
    c. Increase the number of coils
    d. Decrease the amount of current

Properties of Electricity

1. Electrical current is measured in
    a. Amps
    b. Volts
    c. Ohms
    d. Watts

2. Voltage is best described as the _____ in the circuit.
    a. Electron flow
    b. Resistance
    c. Pressure
    d. Current

3. Voltage can only be measured when there is a current.
    ☐ True
    ☐ False

4. Voltages lower than _____ volts are classified as nonhazardous.
    a. 10
    b. 25
    c. 50
    d. 120

5. A 4-AWG wire is _____ than a 12-AWG wire.
    a. Larger
    b. Smaller

6. The two things that must be present for current flow are _____ and _____.
    a. Resistance and wire
    b. Heat and insulation
    c. Voltage and complete path
    d. Tools and fuses

7. The "let go" threshold for current is _____.
    a. 5 mA
    b. 10 mA
    c. 25 mA
    d. 30 mA

8. Another term for watts is _____.
    a. Energy
    b. Heat
    c. Light
    d. Resistance

9. Resistance is measured in _____.
    a. Volts
    b. Amperes
    c. Watts
    d. Ohms

10. Thicker wires have more electrical resistance than thinner wires.
    ☐ True
    ☐ False

11. Match the electrical terms to their hydraulic counterparts.
    a. Voltage   1. Flow
    b. Battery   2. Pressure
    c. Switch   3. Friction/Head Loss
    d. Resistance   4. Valve
    e. Current   5. Pump

Relationships Between Properties of Electricity

1. Ohm's Law is the relationship of voltage, current, and _____.
    a. Resistance
    b. Watts
    c. Magnetism
    d. Heat

2. The current in a circuit is 2 A with a resistance of 4 Ω. What is the voltage?
    a. 6 V
    b. 8 V
    c. 10 V
    d. 16 V

3. How many amps are there in a circuit with 1200 W and 120 V?
    a. 10 A
    b. 30 A
    c. 120 A
    d. 240 A

4. A 40-hp motor operates full time for 1 year. If electricity costs $0.065 kWh, how much did it cost to operate the motor? Assume no surcharges apply.
    a. $16 991
    b. $17 036
    c. $18 995
    d. $22 776

Parallel and Series Circuits

1. The amount of current is the same through any component in a parallel circuit.
   - ☐ True
   - ☐ False

2. A 12-V circuit is wired with three resistors in series. The resistors are rated for 10 ohms (Ω), 5 Ω, and 3 Ω. Find the total equivalent resistance for the circuit.
   - a. 8 Ω
   - b. 12 Ω
   - c. 18 Ω
   - d. 22 Ω

3. The sum of the voltage drops across each load in a series circuit are equal to the voltage supplied to the circuit.
   - ☐ True
   - ☐ False

4. A 12-V circuit is wired with three loads in parallel. The loads are rated for 10 Ω, 5 Ω, and 3 Ω. What is the total current through the circuit?
   - a. 7.6 A
   - b. 12.3 A
   - c. 18.0 A
   - d. 22.8 A

5. Voltage is equal across all components in a parallel circuit.
   - ☐ True
   - ☐ False

Direct Current and Alternating Current

1. One characteristic of DC is
   - a. Fluctuating voltage
   - b. Unidirectional
   - c. 50 or 60 Hz
   - d. Available from wall outlets

2. A solenoid is another name for a
   - a. Coil of wire
   - b. Transformer
   - c. Valve
   - d. Electromagnet

3. This term describes the frequency of cycles in alternating current:
   - a. Hertz
   - b. Voltage
   - c. Ohms
   - d. Amplitude

4. Individual electrons move through electrical circuits and return to their starting point.
   - ☐ True
   - ☐ False

5. In the United States, grounded conductors (wires) in direct current electrical circuits are typically
   - a. Red
   - b. Black or blue
   - c. White or grey
   - d. Yellow

6. This indicates that an extension cord contains an equipment grounding conductor (wire):
   a. Two-prong plug
   b. Yellow stripe on green
   c. Grey or orange exterior
   d. Three-prong plug

7. Equipment grounding conductors (wires) are typically used to return electrons back to the power source and complete the circuit.
   ☐ True
   ☐ False

8. In single-phase AC circuits, either the red or the black wire could be carrying power from the source to the load.
   ☐ True
   ☐ False

9. Most industrial equipment is powered with
   a. Single-phase electricity
   b. Three-phase electricity

10. Three-phase electricity consists of multiple power sources shifted in time relative to one another.
    ☐ True
    ☐ False

11. One advantage of three-phase power is
    a. More expensive to generate compared to single-phase power
    b. More reliable compared to single-phase power
    c. Less variation in voltage compared to single-phase power
    d. Less likely to cause electrical fires than single-phase power

12. How many wires does an ungrounded three-phase AC circuit contain?
    a. 2
    b. 4
    c. 6
    d. 8

Transformers and Short Circuits and Ground Faults

1. Direct current turns off and on 120 times per second.
   ☐ True
   ☐ False

2. A transformer has a 4-to-1 turns ratio, primary to secondary. If the primary voltage is 480 V, what is the voltage of the secondary?
   a. 480 V
   b. 240 V
   c. 120 V
   d. 24 V

3. A magnetic field induces voltage into a conductor moving through it.
   ☐ True
   ☐ False

4. There are 75 000 W on the primary of a transformer with a 4-to-1 turns ratio. How many watts are on the secondary?
   a. 18 750
   b. 37 500
   c. 25 000
   d. 75 000

5. Both a short circuit and an overload create a surge in current that can trip a breaker.
   ☐ True
   ☐ False

6. The circuit breaker tripped. What should the operator do?
   a. Reset the breaker as soon as possible.
   b. Call a qualified electrician.
   c. Replace the circuit breaker.
   d. Bypass the circuit breaker.

Motors

1. The larger motors found in WRRFs typically operate on
   a. Direct current from batteries
   b. Single-phase 110-VAC current
   c. Three-phase 240-VAC current
   d. Three-phase 480-VAC current

2. The permanent magnets of a DC motor are typically located in the
   a. Stator
   b. Squirrel cage
   c. Armature
   d. Commutator

3. Contact between the windings and the commutator are maintained by
   a. Magnetic field
   b. Brushes
   c. Armature glue
   d. Grease

4. In a DC motor, the commutator ensures that only one set of windings is energized at a time.
   ☐ True
   ☐ False

5. What causes the rotor to rotate in an electric motor?
   a. Squirrels running in the cage
   b. Repulsion between two magnetic fields
   c. Electrification of the magnets
   d. Stack tooth interaction with windings

6. In AC motors, the permanent magnets of a DC motor are replaced by
   a. Transformers
   b. Armature stack
   c. Second set of windings
   d. Commutator bar

7. The power formula for a three-phase AC motor includes a factor of 1.762 to compensate for having power from three distinct sources.
   ☐ True
   ☐ False

8. This is the most common type of induction motor:
   a. Synchronous
   b. Squirrel cage
   c. Wound rotor
   d. Single phase

9. Which of the following AC motors will have the slowest synchronous speed?
    a. 2 pole
    b. 4 pole
    c. 6 pole
    d. 8 pole

10. For a squirrel cage motor to produce torque, the operating speed and synchronous speed must be equal.
    ☐ True
    ☐ False

Nameplate Information

1. A motor may operate at voltages that are up to _____ different than its design voltage.
    a. 5%
    b. 10%
    c. 15%
    d. 20%

2. This term describes the number of amps a motor will draw when fully loaded.
    a. Locked rotor amperage
    b. Free spinning amperage
    c. Full load amperage
    d. Base load amperage

3. The amp draw of a motor is highest when it reaches its typical operating speed.
    ☐ True
    ☐ False

4. Calculate the motor speed for a synchronous motor operating at 50 Hz.
    a. 2400 rpm
    b. 3000 rpm
    c. 3600 rpm
    d. 4200 rpm

5. Two motors with the same frame number
    a. Have the same shaft size
    b. Operate at the same revolutions per minute
    c. Draw the same amperes
    d. Generate the same kilowatts (horsepower)

6. A motor with a service factor of 1.1 may
    a. Require extra time to come to full speed
    b. Operate at 110% of its capacity without harm
    c. Run on both 50 and 60 Hz power
    d. Only be used with DC power

7. A motor code for inrush current indicates the amp draw of a locked rotor.
    ☐ True
    ☐ False

8. Motors with this NEMA designation are well suited for use with blowers and pumps:
    a. A
    b. B
    c. C
    d. D

9. **This type of motor should be used in places where flammable gases might accumulate:**
    a. TEFC
    b. TEBC
    c. ODP
    d. TEXP

Motor Control and Routine Motor Operation and Maintenance

1. **The motor FLA nameplate information is for _____.**
    a. Sizing the overloads
    b. Sizing the fuses
    c. Sizing the circuit breaker
    d. Sizing the wire

2. **The frame number is _____.**
    a. The same for every motor
    b. For sizing the fuses
    c. For sizing the circuit breaker
    d. For determining the shaft height

3. **The abbreviation rpm means _____.**
    a. Routine protection maintenance
    b. Revolutions per minute
    c. Rotation per month
    d. Revolving parts maintained

4. **Service Factor is _____.**
    a. The location for the motor
    b. The type of insulation for the motor
    c. The amount that the motor can be overloaded
    d. The type of enclosure on the motor

5. **The work a motor can do is measured in _____.**
    a. Horsepower
    b. Amperes
    c. Voltage
    d. Resistance

6. **Who may take measurements for voltage, amperage, and other electrical parameters?**
    a. Operators
    b. Managers
    c. Trained personnel
    d. Only qualified electricians

7. **If the voltages between the phases in three-phase power are unbalanced**
    a. Motor temperature will increase.
    b. The motor won't start.
    c. Amp draw will increase.
    d. Inspect the windings.

8. **A voltage imbalance may be caused by a problem with the motor controls.**
    ☐ True
    ☐ False

9. When testing a three-phase motor for a voltage or amperage imbalance, it isn't necessary to check all three leads.
   ☐ True
   ☐ False

10. After performing routine maintenance on a three-phase motor, the motor is restarted. Before maintenance was done, the motor was spinning clockwise. Now, it spins counterclockwise. How can this problem be corrected?
    a. Rebalance voltages to within 1%
    b. Switch any two electrical leads
    c. Call the power company
    d. Demagnetize the coils

11. When using an megohmmeter to check the resistance of insulation in a motor:
    a. Check the wattage before testing
    b. Ensure that the motor is connected to the breaker
    c. Isolate the motor from all power sources
    d. Verify that the rotor can spin freely

12. Keep these clean to ensure good airflow and help keep a motor running in the correct temperature range:
    a. Vents
    b. Bearings
    c. Windings
    d. Reservoirs

13. The number one cause of bearing failure is
    a. Dust
    b. Scoring
    c. Overgreasing
    d. Moisture

14. A well-cared-for bearing should last for at least _____ of operation.
    a. 5000 hours
    b. 25 000 hours
    c. 50 000 hours
    d. 100 000 hours

Electrical Prints and Disconnects and Motors

1. Knowing symbols is important
   a. To understand the flow of electricity
   b. To understand the temperature in the circuit
   c. To understand the weather forecast
   d. To understand how a motor works

2. The two main types of disconnects are
   a. Hot and cold
   b. Color coded and numbered
   c. Circuit breakers and knife switches
   d. Negative and positive

3. The thermal-magnetic circuit breaker is found in
   a. Homes
   b. Commercial buildings
   c. Industry
   d. All of these

4. If a knife switch is mounted upside down, it exposes the operator to
    a. A pinch hazard
    b. A shock hazard
    c. Frostbite
    d. A serious cut

5. Fuses with metal caps on the ends are rated
    a. 25 A and less
    b. 50 A and less
    c. 100 A and less
    d. 60 A and less

Push Buttons and Switches

1. is a symbol for a normally open _____ switch.
    a. Flow
    b. Toilet bowl
    c. Pressure
    d. Float

2. is the symbol for a _____.
    a. Normally open flow switch
    b. Normally closed limit switch
    c. Normally open push button
    d. Normally closed push button

3. is a symbol for a _____.
    a. Normally open foot switch
    b. Normally closed temperature switch
    c. Normally open pressure switch
    d. Normally closed foot switch

4. is a symbol for a _____.
    a. Normally closed limit switch
    b. Held closed limit switch
    c. Normally closed pressure switch
    d. Normally closed temperature switch

5. is a symbol for a _____.
    a. Normally closed push button
    b. Held closed push button
    c. Normally closed selector switch
    d. Selector switch

6. A circle is the symbol for a _____.
   a. Pressure switch
   b. Push button
   c. Coil
   d. Foot switch

Relay Coils and Contacts

1. Relays work because of _____.
   a. Magnetism
   b. Resistance
   c. Heat
   d. Direct current voltage

2. The three types of contactors are _____.
   a. Ice cube relay, contactor, overload
   b. Contactor, motor starter, pump
   c. Motor starter, RF converter, controller
   d. Ice cube relay, contactor, motor starter

3. The ice cube relay contacts _____.
   a. Can carry 100 A
   b. Can carry 60 A
   c. Can carry 10 A
   d. Can't carry current

4. The 8-pin relay _____.
   a. Has 10 sets of contacts
   b. Has 8 sets of contacts
   c. Has 6 sets of contacts
   d. Has 4 sets of contacts

5. Push buttons in a control drawing _____.
   a. Are all NO
   b. Are both NO and NC
   c. Are all NC
   d. There are no push buttons in control circuits.

Safety-Related Work Practices

1. The hazard risk assessment is done by _____.
   a. A qualified person
   b. Anyone working on the equipment
   c. A person with limited training
   d. A risk analysis is not necessary.

2. The arc flash boundary is _____.
   a. 12 in.
   b. 42 in.
   c. Where the energy is 1.2 cal/cm^2
   d. Not important

3. The limited approach boundary is _____.

    a. A shock boundary

    b. 12 in.

    c. An arc flash boundary

    d. Where PPE is required

4. Personal protective equipment for shock protection _____.

    a. Is rated in calories

    b. Is insulated rubber

    c. A boundary set by voltage

    d. Any clothing that does not burn

CHAPTER 7
Pumps and Lift Stations

Classification of Pumps

1. Which of the following is an example of a kinetic or dynamic pump?
    a. Air lift
    b. Centrifugal
    c. Rotary lobe
    d. Positive displacement

2. Which of the following is an example of a positive displacement pump?
    a. Hydraulic ram
    b. Centrifugal
    c. Rotary lobe
    d. Turbine

3. Positive displacement pumps are used to pump slurries that are typically more than 4% to 5% TS.
    ☐ True
    ☐ False

Centrifugal Pumps

1. The theoretical maximum lift for any pump is
    a. 4.5 m (14.7 ft)
    b. 6.2 m (20.2 ft)
    c. 10.3 m (34.0 ft)
    d. 25.3 m (83.0 ft)

2. Centrifugal pumps are used to pump up to what percent solids?
    a. 0.04%
    b. 0.4%
    c. 4%
    d. 14%

3. Wastewater pumps were traditionally designed to pass a maximum solid up to a _____ diameter.
    a. 25 mm (1 in.)
    b. 51 mm (2 in.)
    c. 76 mm (3 in.)
    d. 102 mm (4 in.)

4. The primary reason to locate a pump well below grade adjacent to a wet well is to _____.
    a. Provide additional discharge head
    b. Provide positive suction head
    c. Provide room for a parallel pump
    d. Cool the motor

5. In an extended shaft pump configuration in which the motor is one or more levels above the pump, a long shaft may require
   a. A cooling jacket
   b. An immersible motor
   c. Intermediate bearings
   d. A double mechanical seal

6. Submersible pumps are mostly cooled by
   a. Heat transfer to the piping
   b. Heat transfer to the chains
   c. Heat transfer to the fluid being pumped
   d. An external heat exchanger above grade

7. An immersible pump would be designed for which of these scenarios?
   a. Everyday operation in a wet well
   b. Above-grade fan-cooled operation
   c. An accidental dry pit flood from a pipe failure
   d. Use as a sump pump in a dry pit pump room

8. Which of these WRRF functions would a split case pump potentially fit?
   a. Effluent reuse water pumping
   b. Primary sludge pumping
   c. Scum pumping
   d. Return activated sludge pumping

9. An impeller with a single shroud behind the pump vanes is referred to as what type of impeller?
   a. Grit impeller
   b. Closed impeller
   c. Open impeller
   d. Semiopen impeller

10. Because the vanes are sandwiched between two shrouds, which type of impeller may clog more easily?
    a. Grit impeller
    b. Closed impeller
    c. Open impeller
    d. Semiopen impeller

11. Chopper, cutter, and higher efficiency semi-open impeller pump designs are pump manufacturer responses to handle what modern problem?
    a. Increased coffee ground content of wastewater
    b. Increased grit content of wastewater
    c. Higher temperature wastewater
    d. Stringy solids content like flushable wipes

12. In which of these scenarios would it be prudent to check the rotation of a pump?
    a. After an extended power outage
    b. After no operation of the pump for 2 months because of dry weather
    c. After reinstalling a rebuilt pump
    d. At daylight saving time change (twice per year)

13. Which type of bearing resists movements perpendicular to the shaft?
    a. Radial bearing
    b. Cross bearing
    c. Thrust bearing
    d. Inverted bearing

14. Which type of bearing resists movement of the shaft as the impeller pushes toward the motor?
    a. Radial bearing
    b. Cross bearing
    c. Thrust bearing
    d. Inverted bearing

15. What is the purpose of wear rings in a non-clog closed impeller pump?
    a. Resist thrust loads
    b. Prevent recirculation from discharge to suction
    c. Sequester grit in the casing
    d. Tear or cut flushable wipes

16. Modern versions of semi-open impellers use a special semi-open impeller to handle wipes. The impeller typically has a close tolerance with the
    a. Wear rings
    b. Wear plate
    c. Volute
    d. Back plate

17. When the wear rings or wear plate become worn, what is likely to occur?
    a. Bearing failure
    b. Seal failure
    c. Pump efficiency will decrease
    d. Motor high temperature will occur

18. The cylindrical area where the shaft passes into the pump and packing is provided to seal off the leakage is called what?
    a. Mechanical seal
    b. Bearing chamber
    c. Eye of the impeller
    d. Stuffing box

19. A special segment of packing that allows pressurized water to be distributed through the stuffing box is called what?
    a. Retainer ring
    b. Water ring
    c. Lantern ring
    d. Seal ring

20. At a minimum, it is recommended that seal water be supplied to the stuffing box _____.
    a. At least 10 psi more than the pump suction pressure
    b. At least 10 psi more than the pressure at the force main discharge
    c. At least 5 psi more than the pump suction pressure
    d. At least 5 psi more than the pressure of fluid within the pump

21. If there is water dripping from a pump seal at a couple drips each second, an operator or maintenance technition should immediately do what?
    a. Grab a torque wrench and set the packing gland bolts to recommended torque.
    b. Loosen the packing gland until a steady stream occurs.
    c. Make no adjustment to the packing gland.
    d. Increase the seal water pressure.

22. Which of these items is a critical part of a mechanical seal?
    a. Seal face
    b. Seal fingers
    c. Seal shoulder
    d. Seal foot

23. What is the biggest disadvantage of a mechanical seal?
    a. Space
    b. Cost
    c. Size
    d. Noise

24. Some pumps are configured with a gravity-forced lubrication for the seal. The cylindrical unit that holds the lubricant is often referred to as what?
    a. Seal push
    b. Quench pot
    c. Potted seal
    d. Forced seal

25. Which of these pump/motor configurations eliminates bearings in the pump?
    a. Extended line shaft
    b. Frame mounting
    c. Belt and sheaves
    d. Close coupling

26. Which of these pump/motor configurations allows a pump to operate at a lower speed than the motor without a gear box or variable-frequency drive (VFD)?
    a. Extended line shaft
    b. Frame mounting
    c. Belt and sheaves
    d. Close coupling

27. What is typically laser aligned before starting up a frame-mounted pump and motor?
    a. Pump and motor bearings
    b. Pump and motor shafts
    c. Mechanical seals
    d. Motor support feet

28. What may happen when a pump operates "left of its curve"?
    a. It will run out.
    b. Increased recirculation may occur.
    c. Wear ring gap will decrease.
    d. Bearing speed will double.

29. The situation in which air bubbles form and collapse in the vicinity of the impeller is called _____.
    a. Sedimentation
    b. Fragmentation
    c. Aeration
    d. Cavitation

30. Wire to water efficiency takes into consideration both _____.
    a. Pump efficiency and transformer efficiency
    b. Pump efficiency and wiring efficiency
    c. Pump efficiency and motor efficiency
    d. Motor efficiency and wiring efficiency

Positive Displacement Pumps

1. The two categories of positive displacement pumps are
    a. Rotary and radial
    b. Reciprocating and centrifugal
    c. Reciprocating and rotary
    d. Radial and centrifugal

2. The discharge pressure of a positive displacement pump is limited by the flow through the pump.
    - ☐ True
    - ☐ False

3. The discharge flow from a positive displacement pump is nearly directly proportional to its
    a. Discharge pressure
    b. Speed
    c. Suction head
    d. Oil temperature

4. A good use of a positive displacement pump would be
    a. Influent lift station at a WRRF
    b. Pumping secondary clarifier effluent for reuse water
    c. Pumping 8% TSS thickened primary sludge
    d. Pumping 0.2% TSS centrifuge centrate

Diaphragm and Disc Pumps

1. The up-and-down motion of a double disc diaphragm pump is produced by what type of cam?
    a. Circular
    b. Square
    c. Eccentric
    d. Elliptical

2. What determines direction of flow in most diaphragm pumps?
    a. Direction of motor rotation
    b. Ball check valve orientation
    c. Direction of diaphragm movement
    d. Guide vane orientation

3. The flexible section of a double disc diaphragm pump that seals the pump is called the
    a. Cam
    b. Connecting rod
    c. Disc
    d. Trunnion

4. A double disc pump must have a clack valve (or foot valve) if
    a. It is pumping thin solids
    b. It is pumping thick solids
    c. It has a high discharge head
    d. It needs to provide a suction lift

5. Diaphragm and double disc pumps typically have
    a. Suction, discharge, and reload cycles
    b. Reload, stopped flow, and discharge cycles
    c. Suction and discharge cycles
    d. Four independent cycles

6. Small diaphragm pumps are most typically driven by
    a. Solar power
    b. Hydraulic pumps
    c. Compressed air
    d. Turbines

7. If a diaphragm pump is sequencing (reciprocating) but there is no flow, a likely possibility is
   a. A ball check is stuck
   b. A rod is broken
   c. A leak is present in the downstream line
   d. Power is off

8. For larger diaphragm pumps (>100 mm [4 in.]) with electric motors, the camshaft (or driveshaft) is typically rotating
   a. At more than 1800 rpm
   b. At approximately 1200 rpm
   c. At approximately 900 rpm
   d. At less than 110 rpm

Plunger Pumps

1. While the pump is operating and under load, when is it safe to close or restrict the discharge valve?
   a. Only when pumping water
   b. At any time
   c. Never
   d. Only when a pressure gauge is use

2. What is the purpose of air chambers before and after a plunger pump?
   a. To contain excess solids
   b. To equalize pressure and dampen pulsations
   c. To hold the diaphragm bladder in place
   d. To reduce the friction from pumping thick sludge

3. A broken shear pin in a plunger pump usually indicates
   a. Normal wear is occurring
   b. The main shaft is damaged
   c. The motor has failed
   d. Overpressurization has occurred

4. Solids movement around the plunger from the pumping chamber to outside is limited by
   a. Water pressure
   b. Packing
   c. Mechanical seal
   d. Air pressure

5. Increasing the eccentricity of a plunger pump cam will
   a. Decrease flow
   b. Increase speed
   c. Increase stroke length
   d. Decrease speed

6. Increasing stroke length but keeping the speed of a plunger pump the same will
   a. Result in higher pressure
   b. Increase the flowrate
   c. Decrease the pressure
   d. Result in less flow

7. The most frequent maintenance activity for a plunger pump is typically
   a. Measuring plunger wear
   b. Calibrating speed and flow
   c. Replacing a ball check valve
   d. Lubrication

8. A triplex plunger pump has three plungers and
    a. No motor
    b. One motor
    c. Two motors
    d. Three motors

Piston Pumps

1. Typical maximum discharge pressure for a piston pump is approximately
    a. 7 kPa (1 psi)
    b. 70 kPa (10 psi)
    c. 700 kPa (100 psi)
    d. 7000 kPa (1000 psi)

2. The piston pump control valves, which consist of a disc, shaft, and hydraulic cylinder are called
    a. Ball valves
    b. Plug valves
    c. Gate valves
    d. Poppet valves

3. When piston pumps are pumping dewatered biosolids, it is common to place a screw auger
    a. Immediately after the pump in the discharge piping
    b. On the suction side of the pump
    c. At the end of the force main
    d. Midway through the discharge piping

4. The primary purpose of a water/polymer injection ring on the discharge piping on a pump conveying high solids cake is to
    a. Reduce the density of the sludge
    b. Increase the cake solids
    c. Decrease the friction at the pipe wall
    d. Decrease the flowrate

5. The motive force for operation of a piston pump is derived from a
    a. High-torque gear drive
    b. High-torque screw
    c. Pneumatic diaphragm
    d. Hydraulic oil pumping system

6. The approximate size of piston pumps is
    a. 150 to 300 mm (6 to 12 in.)
    b. 25 to 100 mm (1 to 4 in.)
    c. 200 to 600 mm (8 to 24 in.)
    d. 150 mm (6 in.)

7. When pumping cake sludge, the water or polymer added to the sludge at the discharge of the pump
    a. Does not change the cake solids
    b. Is very small, only approximately 10% of the cake flow
    c. Is very small, only approximately 1% of the cake flow
    d. Increases the cake solids

8. One of the main wear items for a piston cake pump is a/an
    a. Trunnion
    b. Eccentric cam
    c. Clack valve
    d. Poppet valve

Rotary Lobe Pumps

1. Why are external check valves not typically required for rotary lobe pumps?
    a. There is very little pulsation.
    b. The pumps have built-in clack valves.
    c. The tolerance between lobes and between lobes and the wall are very small.
    d. Sludge flows toward high pressure on its own.

2. Rotary lobe pumps generally have how many shafts?
    a. None
    b. One
    c. Two
    d. Three

3. If the rotational speed of a rotary lobe pump goes from 20 to 60 rpm, the flowrate through the pump will increase by approximately how much?
    a. 40%
    b. 80%
    c. 100%
    d. 200%

Progressing Cavity Pumps

1. Run-dry is a condition in which a progressing cavity pump
    a. Operates without oil
    b. Operates without grease
    c. Operates without material pumped through it
    d. Operates during weather without rain

2. A progressing cavity pump only pumps material one direction.
    ☐ True
    ☐ False

3. When a progressing cavity runs dry, the temperature in the pumping unit increases because of
    a. Excessive motor heat
    b. Friction between rotor and stator
    c. Bearing heat
    d. Exothermal chemical reactions

4. The reservoir of glycerin or a similar fluid used to protect the mechanical seal in corrosive or otherwise difficult material is called a
    a. Reserve fluid
    b. Seal guardian
    c. Seal saver
    d. Quench pot

Screw Pumps

1. The flowrate through a screw pump increases with what?
    a. An increase in discharge pressure
    b. A decrease in wet well level
    c. An increase in wet well level
    d. Nothing. It is a constant flow pump.

2. Screw pumps are often found in which of these unit processes?

   a. Primary sludge pumping

   b. Influent pumping

   c. Secondary clarifier effluent pumping.

   d. Internal recycle for total nitrogen removal

Peristaltic Pumps

1. At the pressures encountered in most WRRF applications, the ratio of flow to pump speed for a peristaltic pump will be most nearly what number?

   a. 0.5

   b. 1.0

   c. 1.5

   d. 2.0

2. What compresses the hose in a peristaltic pump?

   a. Rollers or boots

   b. Shoes or shafts

   c. Rollers or shafts

   d. Shoes or rollers

Lift Stations/Pumping Stations

1. One reason to use a lift station is when

   a. Flow needs to go downhill

   b. Gravity sewer would require deep excavation

   c. Manholes take up too much space

   d. Plastic pipe is to be avoided

2. In a collection system, lift stations generally decrease in size as they progress further downstream.

   ☐ True

   ☐ False

3. An extra or spare pump is provided in lift stations for

   a. Head

   b. Redundancy

   c. Flow

   d. Symmetry

4. In case of a main power source failure, lift stations provide capacity using an onsite generator, an engine-driven pump on site, or

   a. Redundant pump

   b. Larger pumps

   c. Portable pump connections

   d. More smaller pumps

5. Which type of lift station has a dedicated pump room?

   a. Submersible lift station

   b. Grinder pumping station

   c. Return activated sludge station

   d. Dry-pit lift station

6. For a submersible pumping station, electrical meters and _____ are typically at grade.
   a. Base elbow
   b. Controls
   c. Pump
   d. Drop pipe

7. In a dry pit pumping station, the pumps are situated in the
   a. Wet well
   b. Control room
   c. Garage
   d. Dry well

8. One of the most important safety concerns in a dry pit station is
   a. Ventilation of the dry well
   b. Controlling seal water leakage
   c. Sufficient light for pump maintenance
   d. Maintaining pipe coatings

9. What is almost always required when entering a wet well?
   a. Hearing protection
   b. Confined space permits/procedures
   c. Stairs
   d. Screens

10. A pump that is installed in a dry well but can operate through a catastrophic flood of the dry well is called a
    a. Survivor pump
    b. Backup pump
    c. Engine pump
    d. Dry-pit submersible pump

11. The nickname for a preconstructed steel dry pit station with ladder access tube and a lower cylindrical area for pumps and controls is
    a. Tube station
    b. Can station
    c. Barrell station
    d. Subterrain dry pit

12. Which of these factors is an advantage of a dry pit station?
    a. Use of manhole steps
    b. Smaller footprint
    c. Shelter from weather during maintenance
    d. Lower cost compared to submersible lift stations

13. Which of these factors is an advantage of a submersible lift station?
    a. Smaller footprint
    b. Higher capital costs
    c. Quick and easy inspection of each pump
    d. Stairs are provided

14. Two of the key design criteria for a typical lift station are flow and
    a. Elevation
    b. Difference in elevation from influent sewer to the top of the wet well
    c. Difference in elevation from the top of the wet well to the influent pipe
    d. Difference in elevation from influent sewer to discharge point

15. In addition to total daily flow, sizing a pumping station includes consideration of

    a. Concrete weight

    b. Peak flows

    c. Concentration of phosphorus

    d. Pipe material

16. In a triplex station, typically

    a. One pump can handle the peak flow

    b. Two pumps can handle the peak flow

    c. Three pumps must run to pump peak flow

    d. Four pumps must run to pump peak flow

17. In a duplex station, each pump should be able to pump

    a. Approximately 1.5 times the peak flow

    b. Half the peak flow

    c. The peak flow

    d. Twice the peak flow

18. When a wet well is designed with a drop from the influent pipe to the water level, what is a possible negative result?

    a. Pump cavitation resulting from excess air in the wet well

    b. Frequent pump starts

    c. Pump cavitation resulting from vortexing (not enough submergence)

    d. Pump clogging

19. The lowest points in a wet well should be near

    a. The influent sewer

    b. The floats

    c. The pump suctions

    d. The access hatch

20. Pump removal and loading is typically simpler with a

    a. Dry pit station with block building

    b. Can station

    c. Dry pit station with fiberglass building

    d. Submersible pumping station

21. Two main ways of protecting a pump from debris are

    a. Pollution prevention and screening

    b. Screening and dissolved air flotation

    c. Size reduction and screening

    d. Screening and vacuuming

22. Automated bar screens typically operate

    a. On a timer

    b. On differential/upstream level

    c. On temperature

    d. On a timer and differential/upstream level

23. It is common to back up bubbler systems, level probes, ultrasonic sensors, and pressure transducers with

    a. Hard drives

    b. Extra pumps

    c. Screening

    d. Float sensors

24. Which of these sensors is located completely above the water level?
    a. Ultrasonic sensor
    b. Bubbler
    c. Pressure transducer
    d. Float sensor

25. SCADA stands for _____ _____ and Data Acquisition
    a. Scum Control
    b. Standard Control
    c. Supervisory Control
    d. Supervisory Care

26. Telemetry systems for communicating from the lift station to a central monitoring point are typically accomplished by leased phone line, cell signal, or
    a. Infrared remotes
    b. Sound waves
    c. Radio waves
    d. Pressure waves

27. The main reason for telemetry system alarms is
    a. To alert neighbors
    b. To minimize operator time at the pumping station
    c. To collect data
    d. To eliminate phone lines

28. A telemetry alarm device preprogrammed to use a traditional phone line to call a list of operator phone numbers when an alarm occurs is a (an)
    a. Cell phone
    b. Chatter box
    c. Autocaller
    d. Autodialer

29. Two types of radio frequency communication commonly used for telemetry communication include licensed frequency and
    a. Dual frequency
    b. Unlicensed frequency
    c. Rapid radio
    d. Broad spectrum radio

30. Downsides to cellular data telemetry systems include
    a. Short life of technology (early obsolescence) and tower damage during storms
    b. Interference from amateur radio and storm damage
    c. Proprietary technology and amateur radio interference
    d. No known issues—they are bulletproof

31. One likely source of odor from lift stations is
    a. Not enough detention time in the lift station
    b. Sunlight/heat on the lift station cover
    c. Long detention time in upstream force mains
    d. High dissolved oxygen in the sewer

32. Nitrate-based odor control primarily
    a. Masks the odor
    b. Provides an alternate energy source for bacteria to use while in the collection system
    c. Provides vitamins for bacteria
    d. Chemically eliminates acids in the wastewater

CHAPTER 8
Aeration Systems

Aeration Theory

1. The largest single expense in a WRRF is typically _____.
   a. Treatment chemicals
   b. Potable water for seal flushing
   c. Odor control
   d. Aeration system energy

2. The oxygen content in air is approximately _____.
   a. 1%
   b. 21%
   c. 78%
   d. 99%

3. How many kilograms (pounds) of oxygen are required to convert 1.00 kg (lb) of ammonia into nitrate?
   a. 1.00
   b. 2.26
   c. 4.57
   d. 7.68

4. The DO saturation point in warm water (86 °F [30 °C]) at sea level is _____ mg/L.
   a. 1.0
   b. 7.6
   c. 14.6
   d. 21.0

5. Oxygen contained within air bubbles injected to a liquid is referred to as DO.
   ☐ True
   ☐ False

6. Small air bubbles have a higher oxygen-transfer efficiency than larger bubbles.
   ☐ True
   ☐ False

7. _____ are used to agitate the water in the process in order to promote the transfer of DO into the liquid.
   a. Air valves
   b. Progressing cavity pumps
   c. DO meters
   d. Mechanical aerators

Types of Aeration Systems

1. Diffused air systems typically have lower capital costs (initial purchase costs) than mechanical aerators.
   ☐ True
   ☐ False

2. Diffused air systems typically provide a higher degree of process control than mechanical aeration systems.
   ☐ True
   ☐ False

Diffused Air Systems

1. Place a check mark next to each of the following items that are components of a diffused aeration system:
   ☐ Blowers
   ☐ Horizontal rotors
   ☐ Diffusers
   ☐ pH analyzers
   ☐ Air valves
   ☐ Controls
   ☐ DO analyzers
   ☐ Air downcomers

Blowers

1. _____ is when low-pressure air is injected below the water surface in a process tank in order to transfer DO into the liquid.
   a. Diffused aeration
   b. Mechanical aeration
   c. Virtual aeration
   d. Theoretical aeration

2. _____ blowers are commonly used at WRRFs.
   a. Magnetic and air foil
   b. Rotating and stationary
   c. Mechanical and diffused
   d. Centrifugal and positive displacement

3. _____ are both types of positive displacement blowers.
   a. Rotary screw and turbo
   b. Turbo and rotary lobe
   c. Rotary lobe and rotary screw
   d. Centrifugal and turbo

4. _____ blowers produce fairly constant airflow rates over a range of different pressure conditions.
   a. Mechanical
   b. Centrifugal
   c. Diffused
   d. Positive displacement

5. _____ blowers produce a fairly wide range of airflow rates over a range of slightly differing pressure conditions.
   a. Mechanical
   b. Centrifugal
   c. Diffused
   d. Positive displacement

6. It is becoming increasingly common that blowers are provided from the manufacturer with complete controls packages.
   ☐ True
   ☐ False

Blower Selection Criteria

1. The _____ is calculated as the sum of the velocity component and the static component.
   a. Blower performance curve
   b. Diffuser OTE
   c. System pressure
   d. Life-cycle cost

2. _____ takes into account the total cost of ownership of a piece of equipment, which can be factored into the decision of whether or not to purchase equipment.
   a. O&M cost factoring
   b. Condition assessment
   c. Life-cycle cost analysis
   d. Asset management

3. The _____ of a blower is the work done by a blower divided by the energy used.
   a. Efficiency
   b. Capacity
   c. Pressure rating
   d. Temperature

4. Hearing protection is an important type of personal protective equipment to wear when working around most blowers.
   ☐ True
   ☐ False

5. The following are important criteria to consider when selecting a new blower for a WRRF. Check all that apply.
   ☐ Maintenance requirements
   ☐ Life-cycle costs
   ☐ Footprint
   ☐ Noise levels
   ☐ Energy efficiency
   ☐ Controls options
   ☐ Performance requirements
   ☐ Aesthetics

6. The length of time that a person is exposed to a noise has little to no effect on long-term hearing damage.
   ☐ True
   ☐ False

7. Monitoring the _____ of a blower may be helpful in detecting a blower problem and correcting it before it causes a failure.
   a. Vibration
   b. Discharge pressure
   c. Airflow rate
   d. Inlet air density

Common Blower Accessories

1. Inlet filters remove dirt and particles from the air entering the blower to protect the blower from damage and to _____.
   a. Protect the diffusers from air-side fouling
   b. Prevent scour in the air distribution piping
   c. Protect DO probes from damage
   d. Avoid loss of cross-sectional area in the air piping

2. **Paper elements for inlet air filters are which of the following? Select all that apply.**
   - ☐ More efficient than wire filter elements
   - ☐ Cheaper to purchase than cloth filter elements
   - ☐ Reusable after washing with water
   - ☐ Less efficient than cloth filter elements
   - ☐ Less durable than cloth filter elements

3. _____ provides early indication that an inlet air filter is becoming plugged before blower performance is negatively affected.
   - a. A temperature indicator
   - b. An oil level gauge
   - c. A differential pressure sensor
   - d. A DO probe

4. **Rubber couplings are used to connect blower inlets and outlets to piping to _____. Check all that apply.**
   - ☐ Prevent heat from traveling through the air piping
   - ☐ Prevent blower vibration from traveling through the air piping
   - ☐ Prevent electric current in the blower from traveling through the air piping
   - ☐ Eliminate the need to align the blower connections with the piping
   - ☐ Make it easier to quickly switch out the blowers as needed

5. _____ are commonly used to connect straight-line shafts of multistage centrifugal blowers and their motors.
   - a. Flanged piping
   - b. Mechanical seals
   - c. Flexible couplings
   - d. V-belt sheaves

6. **Motors and positive displacement blowers can be installed side-by-side with their shafts parallel and connected by V-belt sheaves.**
   - ☐ True
   - ☐ False

7. The purpose of a _____ is to decrease the amount of in-rush electricity that is required when a blower is put into operation.
   - a. Transformer
   - b. Motor control center
   - c. Disconnect switch
   - d. Soft starter

8. A _____ is required at every point where a load must be carried between a stationary and moving part.
   - a. Bearing
   - b. Mechanical coupling
   - c. Impeller
   - d. Isolation valve

9. A _____ converts incoming power to an adjustable frequency that can be used to control the speed of the piece of equipment that it is powering.
   - a. Soft starter
   - b. VFD
   - c. Motor control center
   - d. Transmitter

10. **Compared to regular motors, permanent magnet motors have the advantages of being able to achieve higher _____. Check all that apply.**
    - ☐ Efficiencies
    - ☐ Turndown ratios
    - ☐ Speeds
    - ☐ Temperatures
    - ☐ Frequencies

11. A VFD has _____.
   a. The same functionality as a soft starter
   b. The ability to adjust operating speed as well as perform soft starts
   c. The ability only to adjust operating speed, but cannot affect starting and stopping
   d. The ability to affect only the frequency, not the speed, of equipment

12. A(n) _____ prevents air from traveling backward through blower.
   a. Isolation valve
   b. Mechanical seal
   c. Check valve
   d. Inlet throttling seal

Centrifugal Blowers

1. In centrifugal blowers, airflow and pressure are created by _____.
   a. A hose being compressed by a rotating lobed impeller
   b. Rotating lobes meshing to trap air and move it to the discharge
   c. A diaphragm pulling air in through the inlet and pushing it out through the discharge
   d. An impeller rotating and flinging air to the outside of the casing

2. A(n) _____ is used to slow the air and to build pressure in a centrifugal blower.
   a. Impeller
   b. Cutoff
   c. Diffuser plate
   d. Inlet vane

3. In a multistage centrifugal blower, the _____ as the air travels through each stage.
   a. Air flowrate increases
   b. Temperature decreases
   c. Pressure increases
   d. Blower efficiency increases

4. A single-stage centrifugal blower has one _____.
   a. Impeller
   b. Motor
   c. Inlet air filter
   d. Mechanical coupling

5. The characteristics of the inlet air, such as temperature and relative humidity, have very little effect on the performance of a centrifugal blower.
   ☐ True
   ☐ False

6. The Combined Gas law explains the relationship that when a gas is compressed by a blower, the temperature of the gas will _____.
   a. Decrease
   b. Increase
   c. Stay the same

7. Which of the following conditions might lead to surge in a centrifugal blower? Check all that apply.
   ☐ The inlet valve is throttled too much.
   ☐ The discharge air piping is hit by a truck and broken outside the blower building.
   ☐ The discharge isolation valve is closed by mistake.
   ☐ The diffusers are severely fouled.

8. _____ is a blower condition in which the discharge airflow rate is very high and the discharge pressure is very low.
    a. Surge
    b. Cavitation
    c. Choke
    d. Overload

9. The most efficient way to control the output airflow rate of a centrifugal blower is to _____.
    a. Throttle the inlet valve
    b. Adjust the blower speed with a VFD
    c. Adjust the blower speed with a soft starter
    d. Throttle the discharge valve

10. Slowing blower speed creates a new operating curve that is _____ the full speed curve.
    a. Parallel and above
    b. Parallel and below
    c. Perpendicular and to the right of
    d. Perpendicular and to the left of

Integrally Geared Single-Stage Centrifugal Blowers

1. Pressure and airflow rates discharged from integrally geared single-stage blowers are adjusted to meet changing diurnal process requirements _____.
    a. By changing the sheaves
    b. By changing the motors
    c. By adjusting the IGV and/or the DDV
    d. By adjusting the biochemical oxygen demand and/or the total suspended solids

2. Under the right conditions, integrally geared single-stage centrifugal blowers can achieve the highest efficiency of all blowers used at WRRFs.
    ☐ True
    ☐ False

Multistage Centrifugal Blowers

1. One deficiency of multistage blowers as compared to other types of blowers commonly used at WRRFs is _____.
    a. Excessive noise production
    b. Lower design point efficiency
    c. Complicated preventive maintenance task recommendations
    d. Limited turndown capacity

2. The discharge pressure of a multistage centrifugal blower is increased by _____.
    a. Adding stages to the blower
    b. Increasing the diameter of the impeller
    c. Changing from single point to dual point control
    d. Adjusting the opening at the inlet vane

High-Speed-Driven (Turbo) Blowers

1. In a turbo blower system, the blower core refers to the _____.
    a. Cast aluminum impeller
    b. Impeller and the blower scroll (or casing)
    c. Impeller, scroll, motor, and bearings
    d. Impeller, scroll, motor, bearings, VFD, controls, and soundproof enclosure

2. High-speed turbo blowers use bearings that have no direct contact during full-speed operation.
    ☐ True
    ☐ False

3. _____ bearings create a pressurized "air cushion" from the rotational velocity of the shaft within the bearing.
    a. Pillow block
    b. Air foil
    c. Magnetic bearings
    d. Self-aligning thrust

4. The following are benefits offered by purchasing a plug-and-play equipment package:
    ☐ Single-source responsibility
    ☐ Factory testing of system
    ☐ Site-specific customization
    ☐ Simplified installation
    ☐ Reduced maintenance
    ☐ Simplified startup

5. Turbo blowers have a lower peak efficiency than most other types of blowers used at WRRFs.
    ☐ True
    ☐ False

6. _____ motors are often used with turbo blowers in order to meet the blower speed requirements and to provide improved efficiency.
    a. Squirrel cage
    b. Two pole induction
    c. Permanent magnet
    d. Direct current shunt

7. Harmonic filters can be used to eliminate _____.
    a. Noise generated by turbo blowers
    b. Fine particles from the blower inlet air
    c. Abrasive particles from the blower oil system
    d. Induced harmonics from VFDs

Positive Displacement Blowers

1. Rotary lobe blowers are also known as "constant pressure" machines.
    ☐ True
    ☐ False

2. _____ does not occur within a rotary lobe blower, but instead occurs at the blower discharge.
    a. DO transfer
    b. Electricity use
    c. Air velocity increase
    d. Air compression

3. A preventive maintenance task that is commonly needed on rotary lobe blowers is _____.
    a. V-belt inspection and re-tensioning
    b. Air foil bearing inspection
    c. Flexible coupling replacement
    d. Oil pump pressure adjustment

4. Rotary lobe blowers have fairly constant efficiency over a range of discharge pressures.
    ☐ True
    ☐ False

Hybrid Rotary Screw Blowers

1. Rotary screw blowers are considered "hybrid" blowers because they use both _____.
    a. Positive displacement and internal compression
    b. Standard bearings and air bearings
    c. Alternating current power and direct current power
    d. Centrifugal motion and diaphragm displacement

2. Rotary screw blowers use gear drives to increase the speed of the blower above the speed of the motor.
    ☐ True
    ☐ False

Air Distribution Piping and Valves

1. _____ valves are the most commonly used valves for air isolation and control.
    a. Butterfly
    b. Iris
    c. Jet
    d. Turbo

2. Properly placed expansion joints are important in an air distribution system to _____.
    a. Diffuse air into the process tank
    b. Carry heat to the microorganisms in the process tank
    c. Relieve pipe strain caused by thermal expansion
    d. Connect the diffusers to the air grid

3. Automatic air valve actuators should have stops set to only allow travel in the mid-range of the valve.
    ☐ True
    ☐ False

4. Valves on air distribution piping are commonly used to _____. Check all that apply.
    ☐ Create pressure differential in the air distribution system
    ☐ Take an air grid out of service
    ☐ Restrict air entering the blower
    ☐ Control the quantity of air going to a biological reactor

Diffusers

1. The diffuser is the point at which _____.
    a. Waste enters the process tank
    b. Air is injected to the process tank
    c. Inlet air is distributed through the stages of the blower
    d. The DO in the process tank is measured

2. Large bubble diffusers require significantly _____ than fine bubble diffusers.
    a. Less energy for DO transfer
    b. Less maintenance and cleaning
    c. Lower airflow from blowers
    d. More footprint for installation

3. Fine bubble diffusers create bubbles that are _____ mm in diameter.
    a. 0.1–1.0
    b. 1.0–3.0
    c. 3.0–10.0
    d. 5.0–12.0

4. _____ can occur when blower inlet air filters allow too many particles to pass through.
   a. Inlet filter clogging
   b. Diffuser biofouling
   c. Air-side diffuser fouling
   d. DO analyzer clogging

5. _____ is a frequent mini-cleaning technique in which high airflows are sent through membrane diffusers for short periods of time.
   a. Acid washing
   b. Blower pressure trend analysis
   c. Diffuser backwash
   d. Air bumping

6. Treatment processes with _____ typically experience slower diffuser biofouling.
   a. High industrial waste loads
   b. Small septage contributions
   c. High SRTs
   d. Low effluent total suspended solids limits

7. Air transfer efficiency improves after cleaning because of _____.
   a. Increased air velocity and headloss
   b. Smaller bubble size and slower rise rates
   c. Higher solids concentrations and less airflow
   d. Larger bubble size and lower temperatures

Aeration System Control

1. Process control systems can use _____ analyzer measurements to determine whether more or less airflow is needed in a certain process tank or zone.
   a. pH and total suspended solids
   b. Temperature and pH
   c. NH_4-N and DO
   d. DO and total suspended solids

2. A variable pressure aeration control strategy uses the pressure resulting from the _____ to adjust the blower capacity and minimize system pressure.
   a. Least loaded aeration tank
   b. Most open valve
   c. Lowest speed blower
   d. Most stable DO reading

3. Programmable logic controller-based control systems improve aeration system efficiency because _____.
   a. Facility flows and loads are always changing
   b. OTEs are different for various types of diffusers
   c. Alpha factors vary for different aeration systems
   d. Aeration tank configurations have not been standardized

Jet Aeration Systems

1. A jet aeration system produces air bubbles that are comparable in size to those produced by fine pore diffusers.
   ☐ True
   ☐ False

2. An aspect of jet aeration system design that makes maintenance easier is _____.
   a. No moving parts are located inside the process tanks
   b. Only maintenance-free equipment is used
   c. No preventive maintenance tasks are required on any of the system equipment
   d. The process has regularly scheduled shutdowns when maintenance can be done

Surface Aerators

1. High-speed surface aerators _____.
   a. Have the impeller direct-coupled to the motor
   b. Aspirate air down into the process liquid
   c. Use drives to control the speed of the impeller
   d. Achieve the highest standard OTE

2. Typical preventive maintenance tasks for low-speed surface aerators may include _____. Check all that apply.
   ☐ Change the oil in the drive
   ☐ Grease the motor
   ☐ Clean the inlet filters
   ☐ Adjust tension on the V-belt

Horizontal Rotors

1. Operators can adjust the _____ to increase/decrease the OTR of a horizontal rotor.
   a. Detention time in the basin
   b. Temperature of the process water
   c. Speed of the shaft rotation
   d. Elevation of the process tank

2. Decreasing the depth of a horizontal rotor's blade submergence will _____.
   a. Directly cause a near-term failure of the shaft bearings
   b. Shorten the useful life of the unit
   c. Decrease the amount of power needed to operate the horizontal rotor
   d. Increase the OTE

Mixer Aerators

1. Mixer aerators transfer DO into process tanks primarily by _____.
   a. Creating turbulence at the water surface
   b. Drawing air down into the tank through the draft tube
   c. Injecting small bubbles of air into the process fluid
   d. Shearing bubbles from the sparger ring and mixing them into the process liquid

2. Select all of the following that are components of mixer aerators.
   ☐ Submersible motors
   ☐ Submerged mixer body
   ☐ Sparger ring
   ☐ Separate air supply
   ☐ Platform-mounted motor
   ☐ Draft tube

3. Mixer aerators not only provide good OTE, but also provide _____.
   a. Extra mixing energy along the tank floor
   b. High-speed mixing for better process efficiency
   c. Integral DO measurement and control
   d. Heat transfer to the process from the mixer shaft

4. Preventive maintenance tasks for mixer aerators include _____.
   a. Replacing the mechanical coupling when it fails
   b. Changing the oil after the recommended amount of run time
   c. Rebuilding the guide support at the bottom of the tank
   d. Replacing the motor after it fails to run

5. Diffuser fouling is a common issue for mixer aerators that can significantly reduce the efficiency of the unit.
   ☐ True
   ☐ False

6. Inlet guide vane adjustment is the most efficient way to control the capacity of a mixer aerator.
   ☐ True
   ☐ False

CHAPTER 9
Laboratory Procedures

Flow Measurement

1. Flumes tend to be self-cleaning, whereas weirs may accumulate material both upstream and downstream of the weir.
   - ☐ True
   - ☐ False

2. Differential-head flow meters may be used with partially full pipes.
   - ☐ True
   - ☐ False

3. Which type of flow measurement device assumes a constant velocity, but a changing area?
   a. Parshall flume
   b. Orifice plate
   c. Magnetic flow meter
   d. Propeller meter

4. Another name for a triangular weir is a
   a. Cipolletti
   b. V-notch
   c. Contracted
   d. Palmer-Bowlus

5. The flowrate over a rectangular weir has decreased. Water is flowing down the face of the weir with no air space between the water and the weir. This change may
   a. Rapidly wear down the face of the weir
   b. Trap additional debris against the weir
   c. Convert the weir to a pressure plate
   d. Affect the accuracy of the flow measurement

6. Where should an ultrasonic level indicator be mounted for a weir or flume?
   a. Directly over the throat
   b. Upstream of the drawdown
   c. Within the divergence section
   d. Immediately behind the weir edge

Sampling

1. Results for samples can be extrapolated to the populations they were taken from assuming the samples are representative.
   - ☐ True
   - ☐ False

2. It is possible to know whether or not a single sample is representative of a population even when no other information about the population is known.
   - ☐ True
   - ☐ False

3. The standard deviation is an indicator of how different the sample results for a population are from the average result.
    ☐ True
    ☐ False

4. Samples should generally be collected from locations where the waste stream is well mixed.
    ☐ True
    ☐ False

5. A set of 30 samples was collected from the influent to a WRRF. The average BOD_5 concentration was 275 mg/L, with a standard deviation of 25 mg/L. Which of the following sample results is the most representative?
    a. 175 mg/L
    b. 225 mg/L
    c. 275 mg/L
    d. 325 mg/L

6. Within a normal bell curve distribution, what percentage of the sample results will be near the average plus or minus one standard deviation?
    a. 50%
    b. 68%
    c. 95%
    d. 99%

7. The liquid treatment side of a WRRF consists of screening, grit removal, primary clarification, secondary treatment, and disinfection. The operator wants to know the average BOD_5 loading to the secondary process. Where should they collect samples for BOD_5?
    a. Influent
    b. Primary effluent
    c. Secondary process
    d. Disinfection

8. While sampling the influent wastewater, the operator notices that there are small chunks of grease and paper floating on the surface of the channel in a few places. Total grease and paper coverage is less than 10% of the channel surface. The operator should
    a. Reschedule sampling for another day.
    b. Include at least 10% surface material in the sample.
    c. Avoid including surface material because it is not representative.
    d. Randomly sample with no effort to include or exclude floatables.

9. Flumes generally make good sampling locations because the wastewater is well mixed at this location and because
    a. Flumes tend to be self-cleaning.
    b. Flow velocities decrease through the flume.
    c. Flumes are always readily accessible.
    d. Screening and grit removal have already taken place.

10. Wastewater should be sampled from a channel at this location:
    a. Center of the channel at a depth of 40% to 60%
    b. One-third of the way across the channel and near the bottom
    c. Center of the channel and close to the surface
    d. One-third of the way across the channel at a depth of 40% to 60%

Purposes of Sampling

1. Purposes and functions associated with sampling include regulatory compliance, process control and troubleshooting, and evaluation of long-term trends.
    ☐ True
    ☐ False

2. A laboratory result reported as less than 2 mg/L is the same as zero.
   ☐ True
   ☐ False

3. All of the sampling and analyses needed for process control are listed in the NPDES or SPDES permit.
   ☐ True
   ☐ False

4. Care must be taken when comparing sample results from one time period to another time period because facility operating conditions may have changed.
   ☐ True
   ☐ False

5. This document lists the most current requirements for sample preservation and holding times as well as approved testing methods.
   a. *Standard Methods for the Examination of Water and Wastewater*
   b. *Code of Federal Regulations,* Title 40, Part 136
   c. The NPDES or SPDES permit for the WRRF
   d. 503 Regulations

6. In the past year, the WRRF began adding ferric chloride to the influent for odor control. The operator knows that ferric chloride can improve clarifier operation by helping nonsettleable particles stick together. How should the data set be evaluated to determine whether or not chemical addition has improved performance?
   a. Average all of the historic data together
   b. Review only data collected before last year
   c. Review only data collected after chemical addition started
   d. Evaluate the historic data as two groups: before and after addition

Sample and Measurement Collection

1. A grab sample is defined as being collected over a period of 15 minutes or less.
   ☐ True
   ☐ False

2. Flow-proportional samples are typically required for permit compliance with exceptions for parameters like pH, FOG, and VOCs.
   ☐ True
   ☐ False

3. When compositing grab samples together, samples should be poured slowly to ensure accuracy.
   ☐ True
   ☐ False

4. If the average daily flow is unknown, intermittent, or highly variable, flow-proportional samples are not possible.
   ☐ True
   ☐ False

5. An operator collects a single sample by filling a sample bottle at the final effluent. This type of sample is called:
   a. Composite
   b. Discrete
   c. Automatic
   d. Parshall

6. An operator is interested in how the activated sludge process performs under peak load conditions. What type of samples should be collected?
   a. Grab
   b. Timed composite
   c. Flow-proportional composite
   d. Continuous

7. A WRRF discharges to a reservoir capable of holding up to 180 days of effluent before discharge. What type of sample should be collected to determine the average phosphorus concentration in the reservoir?

    a. Grab

    b. Timed composite

    c. Flow-proportional composite

    d. Continuous

8. This type of sample is useful for determining average process performance.

    a. Grab

    b. Composite

    c. Continuous

    d. Automatic

9. Calculate the phosphorus concentration for a time-based composite sample given the following information:

Time	Flow, m³/d	Concentration, mg/L
8:00 a.m.	300	5
10:00 a.m.	450	8
12:00 p.m.	720	9.5
2:00 p.m.	380	6

    a. 6.5

    b. 6.8

    c. 7.1

    d. 7.7

10. Given the following information, determine the sample volume that should be collected at 10:00 a.m. The average daily flow over the past week has been 17.4 ML/d (4.6 mgd). A total of six samples will be collected and the composite sample volume needs to be 3300 mL. The instantaneous flow at 10:00 a.m. is 15.5 ML/d (4.1 mgd).

    a. 270 mL

    b. 380 mL

    c. 490 mL

    d. 600 mL

11. After completing the calculations for flow-paced composite sampling, the operator notices that one of the aliquots will be 87 mL. They should

    a. Increase only that aliquot size to 100 mL.

    b. Verify that the final sample volume will be greater than 100 mL.

    c. Nothing. The final sample volume will be adequate.

    d. Adjust all of the aliquot sizes so the smallest aliquot is at least 100 mL.

12. An autosampler has been programmed to collect a 100-mL sample each time the flow meter measures 4.54 ML/d (1.2 mgd). This type of composite sample is a

    a. Variable frequency composite

    b. Variable volume composite

    c. Simple composite

    d. Continuous composite

Sampling Equipment

1. Sampling equipment should not be constructed from brass, galvanized metal, or other materials that could contaminate the sample.

    ☐ True

    ☐ False

2. All sampling equipment should be rigorously soaked in a dilute acid solution before use.
   - ☐ True
   - ☐ False

3. Generally, sampling devices and containers should be rinsed in the field with the waste stream to be sampled unless they contain preservatives.
   - ☐ True
   - ☐ False

4. Core samplers allow operators to visually inspect layers within a basin, channel, or clarifier.
   - ☐ True
   - ☐ False

5. Powdered gloves have the potential to contaminate samples with
   a. BOD
   b. Zinc
   c. Ammonia
   d. Fecal coliforms

6. Sample faucets must be flushed before sampling for all of these reasons EXCEPT to
   a. Remove accumulated solids
   b. Ensure samples are representative
   c. Increase flow velocity
   d. Maintain proper sampling temperature

7. One advantage of using autosamplers over manual sample collection is
   a. High cost of labor
   b. Compensates for changing conditions
   c. More consistent samples
   d. Unaffected by power outages

8. An autosampler is suspected of contaminating samples with zinc. What can be done to check for contamination?
   a. Replace intake tubing
   b. Remove metal components
   c. Flush sampler with distilled water and analyze
   d. Perform a wipe test of the carboy

9. This may be used to clean autosampler intake tubing to remove biofilm, algae, and other deposits.
   a. Bottle brush
   b. Bleach solution
   c. Pipe cleaners
   d. WD-40

10. Autosampler intake lines should be vertical or sloped and should not contain loops or dips for this reason:
    a. Decreases the likelihood of the line clogging
    b. Ensures gravity drainage between sampling events
    c. Avoids obstructing flow measurement devices
    d. Prevents sample contamination by rising water levels

11. To ensure that the autosampler intake remains at the correct depth within an influent channel, the clear, flexible tubing should be
    a. Housed within a rigid sleeve
    b. Allowed to dangle freely into the channel
    c. Looped over the railing and secured
    d. Weighted with a brick or lead ring

12. Autosampler intakes should be checked regularly when in use to ensure that
    a. The intake is securely bolted to the channel wall
    b. A 0.2-m (8-in.) loop is maintained in the sample line
    c. The internal temperature does not fall below 20 °C (68 °F)
    d. Rags and other debris have not accumulated on the intake

Sample Handling

1. Sample containers should be cleaned before each use.
    ☐ True
    ☐ False

2. Chlorine residual analysis may be performed on a 24-hour composite sample for compliance purposes.
    ☐ True
    ☐ False

3. Regardless of hold time, samples should be analyzed as soon as possible so their characteristics don't change.
    ☐ True
    ☐ False

4. The COC is a legal document that helps maintain the integrity of the sample.
    ☐ True
    ☐ False

5. Generally, washing sampling containers with laboratory-grade detergent followed by a distilled water rinse is adequate for most parameters. What additional cleaning step is needed when sampling for trace metals?
    a. Solvent rinse
    b. Chamois wipe
    c. Acid washing
    d. Heat treatment

6. The hold time for composite samples begins when
    a. The first sample in the composite is taken
    b. The individual grab samples are composited
    c. The last sample in the composite is taken
    d. The sample has been delivered to the laboratory

7. If a laboratory analysis cannot begin right away, samples may be cooled and preservatives may be added to
    a. Increase biological activity
    b. Dissolve solid material
    c. Chelate heavy metals
    d. Minimize biological and chemical changes

8. Biochemical oxygen demand analysis must begin within _____ of sample collection.
    a. 24 hours
    b. 48 hours
    c. 7 days
    d. 28 days

9. Samples for total coliforms, fecal coliforms, and *Escherichia coli* must be analyzed within _____ of collection.
    a. 6 hours
    b. 24 hours
    c. 48 hours
    d. 72 hours

Quality Assurance and Quality Control

1. Data that are consistently biased low or high can still be useful for evaluating trends.
   ☐ True
   ☐ False

2. When using pre-labeled bottles, the blank should always go into bottle number 1.
   ☐ True
   ☐ False

3. The results for samples and their field duplicates are averaged together for reporting purposes.
   ☐ True
   ☐ False

4. Match the quality control sample type to its purpose

   a. Blanks   1. Accuracy
   b. Standard   2. Contamination
   c. Duplicate   3. Drift
   d. Spike   4. Interference
   e. ICV/CCV   5. Precision

5. The difference between a measured result and the true value is the
   a. Error
   b. Precision
   c. Positive bias
   d. Negative bias

6. A standard solution with a concentration of 100 mg/L is analyzed four times. The results are 60 mg/L, 58 mg/L, 63 mg/L, and 56 mg/L. The results can be described as
   a. Accurate, but not precise
   b. Precise, but not accurate
   c. Both accurate and precise
   d. Neither accurate nor precise

7. This type of quality control sample can be used to determine if the correct reagents were used in the proper amounts.
   a. Blank
   b. Standard
   c. Duplicate
   d. Spike

8. A sample batch consists of _____ or fewer samples analyzed together.
   a. 10
   b. 20
   c. 30
   d. 40

9. A batch of samples is analyzed for low-level mercury. The MDL for low-level mercury is 0.02 µg/L. The field blank contains 0.04 µg/L of mercury. The reagent water blank contains <0.02 µg/L of mercury. Sample results range from <0.02 µg/L up to 0.23 µg/L. What must be true?
   a. The field blank result should be subtracted from all sample results.
   b. Mercury contamination is occurring during analysis.
   c. The reagent water result should be subtracted from the field blank result.
   d. Mercury contamination is occurring during sampling.

10. This type of blank is used to assess the cleanliness of sampling equipment and containers before use.
    a. Trip
    b. Field
    c. Rinsate
    d. Reagent

11. One reason to analyze a filter blank is to determine
    a. If filter material was lost in the sample
    b. If contamination occurred during sampling
    c. If background concentrations need to be subtracted
    d. If the dilution water used in the test is contaminated

12. Blank results should always be
    a. Zero
    b. Less than the MDL
    c. Between the MDL and PQL
    d. Equal to the reporting limit

13. Two samples are collected from the final effluent of a WRRF 5 minutes apart. The pH of the first sample is 7.2 and the pH of the second sample is 7.8. What is the most likely reason for the difference?
    a. First sample was not representative.
    b. Samples were collected at different depths.
    c. Effluent quality was changing during sampling.
    d. pH meter is unstable and drifting.

14. Find the RPD between two sample measurements. The sample result is 15.3 mg/L and the duplicate result is 17.6 mg/L.
    a. -2.3
    b. 2.3
    c. -14
    d. 14

15. Calculate the percent recovery for a BOD standard. The certified true value of the standard is 198 mg/L. The measured result in the BOD test is 176 mg/L.
    a. 11.1%
    b. 76.2%
    c. 88.9%
    d. 112.5%

16. An operator needs to make a 15-mg/L standard from a 1000-mg/L stock solution. Available glassware includes a 50-mL Class B beaker and a 100-mL Class A volumetric flask. Only 25 mL of diluted standard is needed. How much stock solution should be used?
    a. 0.75 mL in the 50-mL beaker
    b. 1.5 mL in the 100-mL flask
    c. 0.75 mL in the 100-mL flask
    d. 1.5 mL in the 50-mL beaker

17. A certified standard has a true value of 100 mg/L. The measured result is only 65 mg/L. All of the following may have caused the low result EXCEPT
    a. Stock standard was contaminated
    b. Reagents are old or not used enough
    c. Laboratory procedure not followed
    d. Standard measured with Class B glassware

18. Which of the following statements regarding LFM samples is true?

    a. Interferences always bias results high.

    b. Diluting the sample can reduce an interference effect.

    c. Percent recoveries as low as 75% are acceptable.

    d. Laboratory-fortified matrices cannot be used with every test method.

19. The original sample result is 18 mg/L. Enough standard solution was added to increase the concentration by 15 mg/L. The spiked sample result is 34.3 mg/L. What is the percent recovery?

    a. 16.3%

    b. 47.5%

    c. 90.6%

    d. 108.7%

20. An operator calibrates the ammonia probe using three standards with concentrations of 1, 10, and 100 mg/L NH_3-N. The first sample result is 120 mg/L NH_3-N. The operator should

    a. Report the result as 120 mg/L NH_3-N.

    b. Dilute the sample 1:1 and reanalyze.

    c. Dilute the sample 1:10 and reanalyze.

    d. Report the result as 100 mg/L NH_3-N.

21. An ion-selective electrode is used to analyze samples for nitrate. There are 15 samples total. How many ICV and CCV samples will be needed at a minimum to ensure that the instrument has not drifted out of calibration during analysis?

    a. 1

    b. 2

    c. 3

    d. 4

22. An ion-selective electrode is used to analyze samples for ammonia. The ICV is within range. Five samples are analyzed followed by a CCV. The CCV is within range. Five more samples are analyzed followed by a second CCV. The second CCV is out of range. The operator should

    a. Rerun the second CCV only and report the results.

    b. Recalibrate the ISE and reanalyze the last five samples.

    c. Make a fresh CCV and rerun all 10 samples.

    d. Make a note on the bench sheet and report the results.

23. Twenty-two samples must be analyzed for nitrate on the same day. How many laboratory duplicates will be required?

    a. 1

    b. 2

    c. 3

    d. 4

24. Eighteen samples for nitrate are analyzed on three different days. How many standards will be needed all together?

    a. 1

    b. 2

    c. 3

    d. 4

Analytical Methods

1. Preprogrammed colorimeters and spectrophotometers must be calibrated with a minimum of three standards each time they are used.

    ☐ True

    ☐ False

2. Different analytes absorb light at different wavelengths.
   ☐ True
   ☐ False

3. Three standards are analyzed using a spectrophotometer. The 1-mg/L standard has an absorbance of 0.050. The 5-mg/L standard has an absorbance of 0.250 and the 10-mg/L standard has an absorbance of 0.500. A sample with an unknown concentration of the same analyte has an absorbance of 0.250. What is the concentration?
   a. 1 mg/L
   b. 3 mg/L
   c. 5 mg/L
   d. 10 mg/L

4. A spectrophotometer is used to analyze one certified standard and one unknown sample. After all the reagents are added, the certified standard is a darker color than the sample. What must be true?
   a. The sample concentration is lower than the standard concentration.
   b. There is an interfering substance in the standard solution.
   c. Both the certified standard and the unknown sample are the same.
   d. There is an interfering substance in the unknown sample.

5. A test method for a colorimeter has a range of 0.05 to 2.0 mg/L. The sample result is 2.5 mg/L. The operator should
   a. Report the result
   b. Dilute and reanalyze
   c. Flag the result
   d. Report as 2.0 mg/L

6. The MDL for a particular phosphorus method is 0.2 mg/L as P. The first sample result is 0.1 mg/L. The operator should
   a. Concentrate the sample and reanalyze
   b. Report the result as 0.1 mg/L as P
   c. Repeat the test with a longer sample cell
   d. Report the result as <0.2 mg/L as P

7. One method of increasing the calibrated range of some colorimeter test methods is to
   a. Dilute all of the standards and samples
   b. Use a sample cell that is shorter in length
   c. Report results higher than the highest standard
   d. Reduce the amount of reagent added

8. Sample cuvettes used in spectrophotometers should be carefully wiped with a soft cloth to remove
   a. Air bubbles
   b. Scratches
   c. Fingerprints
   d. Color

9. Gel standards are useful for
   a. Confirming proper operation of an instrument
   b. Confirming the accuracy of the test method
   c. Verifying that reagents are fresh and uncontaminated
   d. Verifying that the test procedure has been followed

10. This type of quality control sample is used to confirm the accuracy of the test method.
    a. Reagent blank
    b. Gel standard
    c. Duplicate
    d. Liquid standard

pH (Hydrogen Ion Concentration)

1. A sample with a pH of 8.2 is considered to be acidic.
    ☐ True
    ☐ False

2. Grease can interfere with pH measurements by coating the electrode.
    ☐ True
    ☐ False

3. A sample with a pH of 4 contains this concentration of hydrogen ions.
    a. 0.0001 mmol/L
    b. 1000.0 mmol/L
    c. 0.0004 mmol/L
    d. 4000 mmol/L

4. Find the average pH for two samples. The first sample has a pH of 5.3 and the second sample has a pH of 7.1.
    a. 5.21
    b. 5.59
    c. 6.20
    d. 6.83

5. pH buffer solutions
    a. May be saved and used over and over again
    b. Can be preserved with hydrochloric acid
    c. Should be considered single-use solutions
    d. Are used to store all types of pH probes

6. For precise pH measurements, sample results
    a. Should be averaged together
    b. Are diluted with pH 4 buffer solution
    c. Are reported to four decimal places
    d. Must be corrected for temperature

Total Alkalinity of Wastewater and Sludge

1. Alkalinity is expressed as mg/L $CaCO_3$.
    ☐ True
    ☐ False

2. Either sulfuric or hydrochloric acid may be used as the titrant in an alkalinity test.
    ☐ True
    ☐ False

3. What is the Normality of a 2-Molar sulfuric acid (H_2SO_4) solution?
    a. 1 N
    b. 2 N
    c. 3 N
    d. 4 N

4. A wastewater sample is titrated to pH 4.5. What type of alkalinity may be calculated if the volume and normality of the titrant is known?
    a. Phenolphthalein
    b. Calcium carbonate
    c. Total
    d. Hydroxide

5. A sample containing bromo-cresol green-methyl red indicator will be this color when the pH is greater than 4.5.
   a. Blue
   b. Purple
   c. Pink
   d. Red

6. Samples from biological treatment processes should be analyzed for alkalinity as soon as possible after collection because
   a. The alkalinity method requires analyses to be completed within 15 minutes of collection.
   b. Biological nitrification or denitrification can decrease or increase alkalinity.
   c. Alkalinity combines with phosphorus and nitrogen in the sample, becoming inert.
   d. Temperature changes in the sample can bias the result high or low.

7. This type of quality control sample is not used with the alkalinity test.
   a. Blank
   b. Standard
   c. Replicate
   d. Duplicate

8. Calculate the total alkalinity of a sample given the following information: sample volume of 50 mL, 0.02-N sulfuric acid titrant, 5.7 mL of titrant added to reach pH 4.5.
   a. 57 mg/L as $CaCO_3$
   b. 114 mg/L as $CaCO_3$
   c. 178 mg/L as $CaCO_3$
   d. 228 mg/L as $CaCO_3$

Total Suspended Solids (Nonfilterable Residue)

1. Larger sample volumes may be filtered when larger diameter filter disks are used.
   ☐ True
   ☐ False

2. One indication that too much material is accumulating on the filter disk is that it takes longer than a few minutes to filter the sample.
   ☐ True
   ☐ False

3. It is acceptable to leave dry areas on the filter disk when filtering small volumes of sample.
   ☐ True
   ☐ False

4. If more than 200 mg of material accumulates on the filter disk during the TSS test
   a. The filter should be split in two
   b. A smaller filter disk is needed
   c. A water trapping crust may form
   d. Sample volume should be increased

5. This type of filter funnel should not be used for the TSS test:
   a. Magnetic
   b. Glass
   c. Gooch
   d. Buchner

6. When measuring small sample volumes for the TSS test, this type of measuring device should be used:
   a. Beaker
   b. Wide-mouth pipette
   c. Graduated cylinder
   d. Transfer pipette

7. How often should a certified standard be included in the TSS test when analyzing samples that will be reported on the discharge monitoring report?
    a. Once per batch
    b. Once per day
    c. Quarterly
    d. Annually

8. Prewashing filter disks for the TSS test
    a. Decreases filter pore size
    b. Seats the filter in the funnel
    c. Removes loose filter material
    d. Reduces fingerprint grease

9. Total suspended solids filter disks may be handled with
    a. Fingers
    b. Tweezers
    c. Gloves
    d. Chopsticks

10. Filter disks should be dried at _____ when determining TSS.
    a. 95 °C (203 °F)
    b. 104 °C (219.2 °F)
    c. 300 °C (572 °F)
    d. 550 °C (1022 °F)

11. After sample filtration is complete, TSS filter disks must be dried
    a. Until the edges curl
    b. For at least 24 hours
    c. Until a constant weight is reached
    d. At 104 °F (219.2 °F) for 1 hour

12. Given the following information, calculate the TSS concentration in the original sample. Initial filter weight = 0.1134 g, filter weight plus residue after drying = 0.1372 g, sample volume = 75 mL.
    a. 3.17 mg/L
    b. 31.7 mg/L
    c. 317 mg/L
    d. 3170 mg/L

13. A filter blank is analyzed with a batch of samples in the TSS test. The initial filter weight was 0.1234 g and the final filter weight was 0.1225 g. What is the most likely reason the final weight is lower than the initial weight?
    a. Excessive prewashing of filter disk
    b. Inadequate prewashing of filter disk
    c. Material transfer from dirty glassware
    d. Filter not dried long enough

14. Multiple aliquots of the same sample are filtered to determine the best volume of sample to use in future tests. Aliquot sizes of 50, 100, 150, and 200 mL were used. The sample passed easily through the filter for each of the first three volumes. With the 200-mL sample, the operator had to wait more than 15 minutes for all of the sample to pass through the filter. Which sample volume should be used for future tests?
    a. 50 mL
    b. 100 mL
    c. 150 mL
    d. 200 mL

15. An operator filtered a sample of influent for the TSS test. After the sample had completely passed through the filter disk, the operator rinsed the filter with several aliquots of deionized water to
    a. Flush dissolved solids away
    b. Make the filter easier to remove
    c. Evenly distribute the solids
    d. Compact the solids

16. Two aliquots of the same mixed liquor sample were analyzed for TSS. The first result was 2200 mg/L and the second result was 2560 mg/L. Calculate the RPD between the two measurements.
    a. 1.2%
    b. 7.0%
    c. 8.6%
    d. 15.2%

Biochemical Oxygen Demand

1. The BOD test measures the concentrations of individual organic compounds in wastewater samples.
    ☐ True
    ☐ False

2. In the United States, BOD samples are incubated for 5 days in the dark at 20 °C (68 °F).
    ☐ True
    ☐ False

3. Biochemical oxygen demand concentrations can be estimated if the COD concentration has been measured and BOD/COD has been established.
    ☐ True
    ☐ False

4. When nitrification inhibitor is added to samples in the BOD test, results are reported as
    a. BOD_5
    b. UBOD
    c. $CBOD_5$
    d. $CNOD_7$

5. All of the following substances are known to inhibit the growth of microorganisms EXCEPT
    a. Low pH
    b. Residual chlorine
    c. Heavy metals
    d. Nontoxic organics

6. When calibrating DO meters, operators must be sure the meter is correcting for differences in oxygen saturation due to
    a. Temperature
    b. Elevation
    c. CO_2 partition
    d. Salinity

7. Salts are added to the BOD dilution water to
    a. Provide necessary nutrients
    b. Increase DO saturation
    c. Buffer against changes in pH
    d. Reduce ionic interference

8. Some BOD samples are seeded to
   a. Add valuable nutrients
   b. Standardize the microorganism types
   c. Ensure there are enough microorganisms
   d. Rehydrate spore forming bacteria

9. Which seed source is best suited for the BOD test?
   a. Well-mixed domestic influent
   b. Chlorinated effluent
   c. Mixed liquor after shaking
   d. Settled primary clarifier effluent

10. The purpose of the seed blanks in the BOD test is to
    a. Check for contamination
    b. Assess the quality of dilution water
    c. Determine the BOD of the seed
    d. Assess the cleanliness of glassware

11. Enough seed should be added to both the G/GA standard and seeded samples to
    a. Consume 0.6 to 1.0 mg/L of DO
    b. Overcome the effects of prior chlorination
    c. Increase the G/GA result to at least 250 mg/L
    d. Determine whether BOD contamination is present

12. *Standard Methods* (APHA et al., 2017) requires a minimum of _____ dilutions per sample for the BOD test.
    a. 1
    b. 2
    c. 3
    d. 5

13. For BOD sample results to be valid, the sample must consume at least _____ of DO over the duration of the test.
    a. 1 mg/L
    b. 2 mg/L
    c. 4 mg/L
    d. 6 mg/L

14. An operator prepares five dilutions of an influent sample for the BOD test. At the end of the test, all five dilutions have less than 1 mg/L of DO remaining. The operator should
    a. Calculate the BOD result using the largest aliquot.
    b. Increase the amount of seed added to each sample.
    c. Decrease aliquot sizes and rerun the test.
    d. Calculate the BOD result and average the results.

15. Samples should be checked for _____ and _____ before aliquoting for the BOD test.
    a. pH and TSS
    b. Residual chlorine and alkalinity
    c. Alkalinity and temperature
    d. pH and residual chlorine

16. Excess DO should be removed from cold samples by
    a. Warming to 20 °C (68 °F) and shaking
    b. Aerating with compressed air
    c. Adjusting the pH with acid
    d. Cooling to ≤6 °C (≤42.8 °F)

17. Plastic caps are placed over the stoppers in the BOD bottles to

    a.   Stop oxygen from entering the BOD bottle

    b.   Prevent evaporation of the water seal

    c.   Keep the stopper from falling out

    d.   Reduce airborne contamination

18. Biochemical oxygen demand bottles are incubated in a dark incubator

    a.   Because incubators don't have windows

    b.   Because oxygen breaks down in sunlight

    c.   To prevent the growth of algae

    d.   To maintain the temperature at 20 °C $\pm$ 1 °C (68 °F $\pm$ 1.8 °F)

19. Given the following information, calculate BOD. There are three dilutions of the same sample. All have the same started DO concentration of 7 mg/L. The first bottle received 25 mL of sample and had a final DO concentration of 6.5 mg/L. The second bottle received 50 mL of sample and had a final DO concentration of 3.5 mg/L. The third bottle received 150 mL of sample and had a final DO concentration of 0.9 mg/L.

    a.   6 mg/L

    b.   13 mg/L

    c.   17 mg/L

    d.   21 mg/L

20. The seed blank has an average depletion of 0.18 mg/L of DO per milliliter of seed solution. What is the minimum amount of seed solution that should be added to both the standard and all seeded samples?

    a.   3 mL

    b.   4 mL

    c.   5 mL

    d.   6 mL

CHAPTER 10
Chemical Storage, Handling, and Feeding

Purpose and Function

1. A chemical reaction that converts insoluble and unsettleable material into particles that can be separated from water through gravity or filtration is a(n)
   a. Acid-base reaction
   b. Coagulation-flocculation reaction
   c. Oxidation–reduction reaction
   d. Dispersion reaction

2. In an oxidation–reduction reaction, the oxidizer is reduced.
   ☐ True
   ☐ False

3. In a chemical treatment process, the chemical is used to control environmental conditions.
   ☐ True
   ☐ False

Theory of Operation

1. The reliability of chemical dosing to the process begins with
   a. A reliable chemical dosing pump
   b. Sufficient on-site storage
   c. Reliable chemical manufacturing and transport
   d. Stable chemical pricing

2. The amount of chemical stored on-site should be limited only by available space for storage facilities.
   ☐ True
   ☐ False

3. Chemicals may be distributed from the bulk storage tank to smaller tanks near the application point. These tanks are known as
   a. Distribution tanks
   b. Process tanks
   c. Feed tanks
   d. Day tanks

4. *Chemical dosing, metering,* and *feed* are synonymous terms for the controlled introduction of chemical to the process.
   ☐ True
   ☐ False

Design Parameters

1. A WRRF anticipates feeding 2650 L/d (700 gpd) of chemical to a new process. The bulk delivery size is 17 000 L (4500 gal). The minimum amount of on-site storage should be
   a. 17 000 L (4500 gal)
   b. 25 500 L (6750 gal)
   c. 37 000 L (9800 gal)
   d. 51 000 L (13 500 gal)

2. Thermoplastics and elastomers are generally made more resistant by the addition of
   a. Nickel
   b. Phosphorus
   c. Hydrogen
   d. Fluoride

Equipment

1. In relation to the bulk chemical delivery, on-site storage should be sized to accept at least
   a. The largest delivery
   b. 1.5 times the largest delivery
   c. Twice the volume of the largest delivery
   d. Three deliveries per day

2. In general, to guard against interruptions in chemical delivery, enough on-site storage should be provided for a
   a. 1-week supply
   b. 2-week supply
   c. 1-month supply
   d. 2-month supply

3. Spill containment in chemical storage areas should be designed to contain
   a. 50% of the storage volume
   b. 100% of the storage volume
   c. 110% of the storage volume
   d. 1.5 times the storage volume

4. If the volume of stored chemical is less than 379 L (100 gal), spill containment is not required.
   ☐ True
   ☐ False

5. Leak detection in chemical containment areas can be accomplished with
   a. A flood switch
   b. A float switch
   c. Tank level indicator with rate of change
   d. All of the above

6. Chemical transfer and dosing pumps should be located outside the containment area to prevent damage to the equipment in case of a spill.
   ☐ True
   ☐ False

7. Chemical spills should be drained back to the facility influent as quickly as possible to make the chemical area safe again.
   ☐ True
   ☐ False

8. Once chemical has spilled from the storage tank into the containment area, it should be considered contaminated and disposed of.
   ☐ True
   ☐ False

9. Controls for chemical containment sump pumps should be
   a. Located within the containment area
   b. Accessible through the facility SCADA system
   c. Automated
   d. Located at the pump but outside the containment area

10. Emergency showers and eyewash stations must deliver water for a minimum duration of
    a. 2 minutes
    b. 5 minutes
    c. 10 minutes
    d. 15 minutes

11. The tepid water required in emergency eyewash stations and showers should be 16 to 38 °C (60 to 100 °F).
    ☐ True
    ☐ False

12. Environmental control equipment specific to storing hygroscopic chemicals includes
    a. Dehumidifiers
    b. Air conditioners
    c. Deicers
    d. UV protection

13. Emergency showers and eyewash stations must be flushed _____ and specifications verified _____.
    a. Daily, weekly
    b. Weekly, annually
    c. Monthly, annually
    d. Monthly, every 3 years

Receiving Facilities

1. Chemical receiving equipment includes protection for vulnerable equipment along the on-site truck route.
   ☐ True
   ☐ False

2. Three methods of transferring chemical from bulk delivery vehicles are gravity, transfer pump, and
   a. Compressed air
   b. Fans
   c. End dump
   d. Vacuum

3. If a cam-lock style quick-connect fitting is leaking, the cams, groove, or _____ may be worn and need replacement.
   a. Hose
   b. Gasket
   c. Diaphragm
   d. Thread sealant

Chemical Transfer

1. A chemical day tank can only be filled once per day.
   ☐ True
   ☐ False

2. The benefit of a day tank is that
    a. It ensures no more than one day's worth of chemical is dosed.
    b. The chemical off-loading procedure is simplified.
    c. Less chemical is required to effect treatment.
    d. The dosing pump is located closer to the application point.

3. Progressing cavity and rotary lobe chemical transfer pumps are examples of
    a. Diaphragm pumps
    b. Positive displacement pumps
    c. Centrifugal transfer pumps
    d. Air-lift pumps

4. The use of a magnetic drive mechanism eliminates the need for
    a. An electric motor
    b. A drive shaft
    c. A drive shaft penetration in the volute
    d. The impeller

5. Diaphragm pumps are used when centrifugal transfer pumps are too small.
    ☐ True
    ☐ False

6. Automated isolation valves are used in chemical transfer systems to prevent
    a. Overdosing chemical to the process
    b. Overfilling the bulk tank
    c. The day tank from being pumped dry
    d. Siphoning and gravity flow

7. In an automatic mode of chemical transfer operation, the transfer sequence is initiated
    a. By the operator
    b. By the day tank level transmitter
    c. By the high-level cut out switch
    d. By the bulk tank level transmitter

Dry Chemical Systems

1. Bulk storage bins for dry chemical are vented _____.
    a. Through a dust collector
    b. Always in the side of the bin
    c. Through a water spray
    d. Only for air off-loading.

2. Hazards of dry chemical dust include inhalation exposure and explosion risk.
    ☐ True
    ☐ False

3. *Bridging* refers to the stoppage of dry chemical flow in the
    a. Fill pipe
    b. Dust collector filter bags
    c. Conical section of the storage bin
    d. Solution tank fill pipe

4. In a manually filled bag supply chemical bin, the drop-down door and exhaust fan provide
    a. Access to the feed hopper
    b. Cooling for the motor
    c. The most accurate dose control
    d. Dust control

5. Volumetric feeders are more precise than gravimetric feeders.
   - ☐ True
   - ☐ False

6. Gravimetric feeders dose chemical based on
   - a. Time
   - b. Weight
   - c. Volume
   - d. Density

7. A screw feeder is usually used as a _____ feeder, whereas a belt feeder is commonly used as a _____ feeder.
   - a. Positive displacement, passive
   - b. Gravimetric, volumetric
   - c. Dry chemical, liquid chemical
   - d. Volumetric, gravimetric

8. Soluble chemicals form suspensions in the solution tank, whereas insoluble chemicals form slurries.
   - ☐ True
   - ☐ False

9. Eductors can use a flow of air or water to create a vacuum that conveys dry chemical to a solution tank or application point.
   - ☐ True
   - ☐ False

10. The bin isolation gate is closed during filling a bulk storage bin to prevent
    - a. Flooding
    - b. Overdosing
    - c. Underdosing
    - d. Dust generation

11. When monitoring a bulk storage bin filling cycle, air discharge through the pressure-relief valve indicates
    - a. The vent valve is closed.
    - b. The dust control filter bags are clogged.
    - c. The bin has been overfilled.
    - d. The bin isolation gate was left open.

12. Calibrating dry feeders involves measuring the discharge from the feeder over a set time period and results in determining the maximum
    - a. Calibration rate
    - b. Feed rate in mass per time
    - c. Feed rate in volume per time
    - d. Dose needed to achieve treatment goals

Liquid Chemical Systems

1. Bulk liquid chemical storage tanks should include a visual level indication such as a(n)
   - a. Open hatch on the top of the tank
   - b. Sight glass
   - c. Pressure transmitter
   - d. Drain flow meter

2. The vent on a bulk liquid chemical storage tank should be closed when filling the tank to prevent spills.
   - ☐ True
   - ☐ False

3. Chemical drums can be used for supplying chemicals with low usage rates, but IBCs are most appropriate for the lowest usage rates.
    ☐ True
    ☐ False

4. Chemical dosing pumps are typically
    a. Diaphragm pumps
    b. Positive displacement pumps
    c. Centrifugal pumps
    d. Progressing cavity pumps

5. Ball check valves are used in which type of chemical dosing pumps?
    a. Peristaltic pumps
    b. Positive displacement pumps
    c. Diaphragm pumps
    d. Progressing cavity pumps

6. Adjusting the stroke length on a diaphragm dosing pump changes the
    a. Pump speed
    b. Stroke duration
    c. Useful diaphragm area
    d. Volume per stroke

7. Pulsation dampeners are most relevant to which type of chemical dosing pumps?
    a. Peristaltic pumps
    b. Centrifugal pumps
    c. Diaphragm pumps
    d. Progressing cavity pumps

8. The purpose of a backpressure valve is to
    a. Prevent siphoning or gravity flow of chemical through the dosing pump
    b. Relieve pressure if the discharge line becomes blocked
    c. Ensure the dosing pump does not run dry
    d. Prevent damage to the system caused by the pressure of the pump

9. The calibration column should be located as close as possible to the dosing pump.
    ☐ True
    ☐ False

10. A chemical dosing pump is set to deliver a constant flow of 10 gph; however, the calibration column is only marked in mL. A drawdown on the calibration column shows 60 mL pumped in 1 minute. Is the pump overdosing or underdosing?
    a. Overdosing
    b. Underdosing

Gas Chemical Systems

1. The purpose of the fusible plug in gas cylinders is to
    a. Prevent gas from escaping in an emergency
    b. Relieve pressure in the event of fire
    c. Make a positive and leak-free connection to the cylinder
    d. Isolate the gas feeder from the application point

2. Chlorine tank cars, ton containers, and cylinders all provide valves for withdrawing liquid or gas chlorine.
    ☐ True
    ☐ False

3. The gasket used to connect a gas container valve to the gas feed system must be a compatible and approved material and be replaced _____.
   a. Every time the container is changed out
   b. Once for every 100 lb of gas fed
   c. Daily
   d. Weekly

4. Vacuum chemical feed systems have a safety advantage over pressure systems because any gas that leaks from the system will be removed from the area by the vacuum.
   ☐ True
   ☐ False

Chemical Application

1. The primary process concern at the point of chemical application is
   a. Corrosion
   b. Mixing
   c. Ventilation
   d. Temperature

2. The primary maintenance concern at the point of chemical application is
   a. Corrosion
   b. Mixing
   c. Ventilation
   d. Temperature

3. The purpose of applying mixing energy at the point of chemical application is to
   a. Aerate the process water prior to chemical application
   b. Cause the reactants to come into contact with each other
   c. Catalyze the reaction
   d. Suspend any precipitates formed to keep them from settling

4. Match the chemical application equipment to the description:

   a. Injection Quill   1. Splits the chemical flow into multiple streams
   b. Diffuser   2. Can keep the chemical from corroding the interior pipe wall
   c. Injection Mixer   3. Used in a tank or open channel to disperse chemical
   d. Static Mixer   4. Provides in-pipe mixing by increasing the flow turbulence

Process Variables

1. A technical data sheet includes all the information found on an SDS.
   ☐ True
   ☐ False

2. A chemical solution with a specific gravity of 0.96 is 4% more dense than water.
   ☐ True
   ☐ False

3. A chemical solution with a specific gravity of 1.21 has a density of
   a. 1.21 g/mL
   b. 10.1 lb/gal
   c. Both a. and b.
   d. None of the above

4. Which term does not refer to the amount of chemical in a chemical solution?
    a. Purity
    b. % active
    c. % neat chemical
    d. Dry chemical

Calculating the Mass of Active Chemical in Neat Chemical and Element or Ion of Use in the Active Chemical

1. The mass of dry chemical in a 30% (wt/wt) chemical solution with a specific gravity of 1.1 is
    a. 1.1 kg/L
    b. 1.33 kg/L
    c. 0.3 kg/L
    d. 0.33 kg/L

2. Find the % (wt/wt) of the hypochlorite ion, OCl^-, in sodium hypochlorite, $NaOCl$. The atomic weight of $Na = 23, O = 16, Cl = 35.5$.
    a. 43%
    b. 51.5%
    c. 23%
    d. 69%

3. A chemical solution with a specific gravity of 1.21 has a density of
    a. 1.21 g/mL
    b. 10.1 lb/gal
    c. Both a. and b.
    d. None of the above

4. Which term does not refer to the amount of chemical in a chemical solution?
    a. Purity
    b. % active
    c. % neat chemical
    d. Dry chemical

Chemical Dose

1. 1 mg/L is equivalent to 1 ppm because
    a. 1 L of water weighs 1 kg.
    b. 1 mL of water weighs 1 g.
    c. 1 L of water weighs 1 000 000 mg.
    d. All of the above

2. A dose of 1 mg/1 000 000 lb is equivalent to a dose of 1 ppm.
    ☐ True
    ☐ False

3. A 50% active chemical solution with a specific gravity of 1.1 is dosed at 20 mg/L to a flow of 75 ML/d. Calculate the feed rate in liters of neat chemical per day.
    a. 3000 L/d
    b. 2727 L/d
    c. 682 L/d
    d. 1500 L/d

4. A 15% active chemical solution with a specific gravity of 1.1 is dosed at 20 mg/L to a flow of 25 mgd. Calculate the neat chemical feed rate in gallons per day.

  a. 3030 gal/d

  b. 3333 gal/d

  c. 68 gal/d

  d. 500 gal/d

5. A dry chemical feeder is delivering a constant feed of 200 kg/h of pure chemical to a flow of 200 ML/d. Calculate the dose in milligrams per liter.

  a. 1 mg/L

  b. 24 mg/L

  c. 48 mg/L

  d. 2400 mg/L

6. A peristaltic chemical dosing pump is delivering a constant feed of 25 gph of 40% active chemical with a product weight of 12 lb/gal to a flow of 90 mgd. Calculate the dose in milligrams per liter.

  a. 0.2 mg/L

  b. 2.7 mg/L

  c. 3.8 mg/L

  d. 6.4 mg/L

Using ABC Equations for Chemical Feed Pump Settings

1. Calculate the chemical feed pump setting in mL/min for dosing 70 mg/L of 40% active chemical with a density of 1.3 kg/L to a flow of 600 ML/d.

  a. 72 917 mL/min

  b. 10 606 mL/min

  c. 29 167 mL/min

  d. 56 090 mL/min

2. A chemical with a density of 1.10 g/cm³ has a specific gravity of 11.0.

  ☐ True

  ☐ False

3. Calculate the chemical feed pump setting in mL/min for dosing 70 mg/L of 40% active chemical with a density of 1.3 kg/L to a flow of 158.5 mgd.

  a. 72 907 mL/min

  b. 10 605 mL/min

  c. 29 163 mL/min

  d. 56 082 mL/min

4. A 20% active chemical solution with a specific gravity of 1.25 is dosed at 100 mg/L to a flow of 100 ML/d. Calculate the neat chemical feed rate in kilograms per day.

  a. 50 kg/d

  b. 500 kg/d

  c. 25 000 kg/d

  d. 50 000 kg/d

5. An 80% active chemical solution with a density of 9.8 lb/gal is dosed at 100 mg/L to a flow of 35 mgd. Calculate the neat chemical feed rate in pounds per day.

  a. 36 488 lb/d

  b. 38 488 lb/d

  c. 40 136 lb/d

Reaction Time

1. Reaction rates always depend on the concentration of the reactants.
   - ☐ True
   - ☐ False

2. A chemical reaction with a higher rate constant will actually proceed more slowly than one with a lower rate constant.
   - ☐ True
   - ☐ False

3. The rate of a first-order reaction depends on the rate constant k and
   a. The initial concentration of the reactant(s)
   b. The concentration of the reactant(s)
   c. Time
   d. All of the above

4. A logarithm and an exponent are inverse functions.
   - ☐ True
   - ☐ False

Side Reactions

1. Given a generic chemical reaction, $A + 2B \leftrightarrow AB_2$, the stoichiometric dose of B needed to react with A is
   a. $2\,g\,B : 1\,g\,A$
   b. $1\,g\,A : 2\,g\,B$
   c. $2\,mol\,B : 1\,mol\,A$
   d. $1\,mol\,A : 1\,mol\,B$

2. Stoichiometric chemical doses are usually sufficient to complete reactions in full-scale treatment processes.
   - ☐ True
   - ☐ False

Process Control

1. The dose needed to achieve treatment goals in a chemical treatment process for any unique water is found
   a. Empirically
   b. Theoretically
   c. Stoichiometrically
   d. From bench tests

2. Variable flows and variable pollutant concentrations in the influent to a WRRF cause
   a. Process upsets
   b. Variable effluent quality in chemical processes
   c. Variable staffing requirements
   d. Variable loading

3. Chemical dosing controls should be programmed to feed an amount of neat chemical that is proportional to the influent flow.
   - ☐ True
   - ☐ False

4. A chemical feeder is set in constant speed to deliver 50 gph of chemical A, 50% (wt/wt) active with a specific gravity of 1.5. The process flowrate varies from a minimum of 10 mgd to a maximum of 60 mgd. What is the minimum and maximum dose of chemical A as active chemical?
   a. 0.63 ppm; 3.8 ppm
   b. 10 ppm; 60 ppm
   c. 15 ppm; 90 ppm
   d. 37.5 ppm; 225 ppm

5. Constant speed control of chemical feeders will produce a consistent effluent quality.
    ☐ True
    ☐ False

6. Variation in the effluent quality over the course of a day can be tolerated if the parameter is evaluated on what type of basis?
    a. Daily maximum
    b. Daily minimum
    c. Instantaneous average
    d. Daily average

7. In a ratio control mode using a feed-back control loop, the chemical dose is adjusted according to an analyzer located
    a. At the unit process effluent
    b. At the unit process influent
    c. At the facility influent
    d. At the facility effluent

Chemical Safety

1. A standard SDS contains _____ sections, _____ of which are non-mandatory.
    a. 12, 0
    b. 16, 0
    c. 12, 4
    d. 16, 4

2. The program that standardizes SDS information and language from international manufacturers is known as the
    a. Hazard Communication Standard
    b. Globally Harmonized System
    c. Chemical Safety System
    d. International Labeling System

3. The signal word in a chemical SDS may be
    a. Either flammable or corrosive
    b. Either severe or moderate
    c. Either fatal or non-fatal
    d. Either danger or warning

4. Match the information contained in an SDS on the left to the appropriate section:

Information	SDS Section
a. LD_{50} values	Identification
b. List of known dangerous reactions	Hazard Identification
c. Emergency number	Composition/Information on Ingredients
d. Chemical formula	First Aid Measures
e. Recommended PPE	Fire Fighting Measures
f. Exposure routes	Accidental Release Measures
g. Pictogram	Handling and Storage
h. Storage conditions	Exposure Controls/Personal Protection
i. Materials and methods of clean up	Physical and Chemical Properties
j. Melting/Freezing point	Stability and Reactivity
k. Suitable extinguishing media	Toxicological Information

5. Which of the following is NOT a form of PPE?

    a. Splash curtain

    b. Safety goggles

    c. Respirator

    d. Face shield

6. Which of the following is NOT an engineering control to minimize exposure?

    a. Ventilation

    b. Flushing ports

    c. Class A exposure suit

    d. Double-contained piping

7. If sufficient risk-minimizing engineering controls are provided, the use of PPE becomes unnecessary.

    ☐ True

    ☐ False

8. A face shield primarily protects against which route of exposure?

    a. Eye

    b. Inhalation

    c. Skin

    d. Ingestion

9. The cartridges used in an air purifying respirator

    a. Are universal

    b. Always remove particulates

    c. Must have an attachment for a supplied air line

    d. Must be matched to the specific chemical hazard

10. A Level A hazardous material protection ensemble is appropriate in any situation involving the risk of chemical exposure.

    ☐ True

    ☐ False

11. Name the hazard identified by the color on a U.S. NFPA placard.

White	
Blue	
Red	
Yellow	

12. The nine classes of hazardous materials identified by DOT correspond to the nine pictograms used by the GHS.

    ☐ True

    ☐ False

13. The right-to-know refers to the idea that workers

    a. Have access to an SDS for any chemical they may be exposed to

    b. Have access to an SDS for chemicals above the reportable quantity

    c. Cannot be unreasonably exposed to chemicals

    d. Must educate themselves on the hazards of chemicals they use

List of Acronyms

%	percent
$(NH_4)2SO_4$	ammonium sulfate
°C	degrees Celsius
°F	degrees Fahrenheit
A	amperes, or amps
AC	alternating current
$Al_2(SO_4)_3$	aluminum sulfate
$AlCl_3$	aluminum chloride
Alk	alkalinity
AOB	ammonia oxidizing bacteria
ASCE	American Society of Civil Engineers
ATAD	autothermal thermophilic aerobic digestion
AWG	American Wire Gauge
AWT	advanced water treatment
BEP	best efficiency point
BFP	belt filter press
BOD	biochemical oxygen demand
BOD_5	5-day biochemical oxygen demand
Btu/scf	British thermal unit per standard cubic foot
Btu	British thermal unit
C	carbon
$C_3H_8O_3$	glycerin
$Ca(OCl)_2$	calcium hypochlorite
$Ca(OH)2$	calcium hydroxide
CaO	calcium oxide
CATC	cyanide amenable to chlorination
CBOD	carbonaceous biochemical oxygen demand
$CBOD_5$	5-day carbonaceous biochemical oxygen demand
CCB	continuing calibration blank
C	concentration; may be shown as concentration 1 (C_1), concentration 2 (C_2), etc.
CCV	continuing calibration verification
CEC	Canadian Electrical Code
CEMA	Conveyor Equipment Manufacturers Association
CEPT	chemically enhanced primary treatment
CFD	computational fluid dynamics
cfm	cubic feet per minute
CFR	Code of Federal Regulations
CFU	colony forming unit
CH_3COOH	acetic acid
CH_3COOOH	peracetic acid
CH_3OH	methanol
CIP	clean in place
Cl_2	chlorine gas
cm	centimeter
CMA	circular mil area
CMMS	computer maintenance management systems
CO_2	carbon dioxide
COC	chain of custody
COD	chemical oxygen demand
cu ft	cubic feet
cu ft/lb	cubic feet per pound
d	day
DAFT	dissolved air flotation thickener
dBA	decibels, A-weighted
DC	direct current
DDV	discharge diffuser vane
DE	drive end
DMR	discharge monitoring report
DMR-QA	discharge monitoring report quality assurance sample
DO	dissolved oxygen
DT	detention time
EBPR	enhanced biological phosphorus removal
E	electromotive force (symbol used in mathematical equations)
EMF	electromotive force
EPAM	emulsion polyacrylamide
ESD	egg-shaped digester
F:M	food-to-microorganism ratio
$Fe_2(SO_4)_3$	ferric sulfate
$FeCl_2$	ferrous chloride
$FeCl_3$	ferric chloride
$FeSO_4$	ferrous sulfate
FLA	full load amps
FOG	fats, oils, and grease
ft	foot or feet
ft/s	feet per second
g	gram
gal	gallon
GBT	gravity belt thickener
gpd	gallons per day
gph	gallons per hour
gpm	gallons per minute
h	hour (International Standard units)
H_2O_2	hydrogen peroxide
H_2SO_4	sulfuric acid
HCl	hydrochloric acid
HDT	hydraulic detention time
HLR	hydraulic loading rate
H-O-A	hand-off-auto
hp	horsepower
hp/1000 cu ft	horsepower per 1000 cubic feet
hp	horsepower
hr	hour (U.S. customary units)
HRT	hydraulic retention time
Hz	Hertz
I	intensity (symbol for current, amps)

IBC	intermediate bulk container		min	minute
ICB	initial calibration blank		MJ	megajoule
ICV	initial calibration verification		mL	milliliter
IDL	instrument detection limit		ML/d	megaliters per day
IDLH	immediately dangerous to life and health		MLSS	mixed liquid suspended solids
IGV	inlet guide vane		MLVSS	mixed liquor volatile suspended solids
in.	inch		mm	millimeter
in. H_2O	inches of water column		mm H_2O	millimeters of water column
ISE	ion-selective electrode		mm^2	square millimeters
J	joule		MW	megawatt
kg	kilograms		MΩ/hp	megaohms per horsepower
kg/d	kilograms per day		$Na_2Al_2O_4$	sodium aluminate
kg/ha	kilograms per hectare		Na_2CO_3	sodium carbonate
kg/h	kilograms per hour		$Na_2S_2O_3$	sodium thiosulfite
kg/L	kilograms per liter		Na_2SO_3	sodium sulfite
kg/m³·d	kilograms per cubic meter per day		$NaHCO_3$	sodium bicarbonate
km	kilometer		$NaHSO_3$	sodium bisulfite
kPa	kilopascal		NaOCl	sodium hypochlorite
kV	kilovolt		NaOH	sodium hydroxide
kVA	kilovolt-ampere		NBS	National Bureau of Standards
kVAR	kilovolt-ampere reactive		NEC	National Electric Code
kW	kilowatt		NEiS	National Electrical Installation Standards
kWh	kilowatt hours		NEMA	National Electrical Manufacturers Association
kΩ	kiloohms		NFPA 70E	National Fire Protection Association Standard for Electrical Safety in the Workplace
L	liter			
L/min	liters per minute		NH_3	anhydrous ammonia
L/s	liters per second		NH_4-N	ammonium nitrogen
lb	pound		NIOSH	National Institute for Occupational Safety and Health
lb/cu ft·d	pounds per cubic foot per day			
lb/d	pounds per day		NO_3	nitrate
lb/gal	pounds per gallon		NOB	nitrite oxidizing bacteria
lb/hr	pounds per hour		NPDES	National Pollutant Discharge Elimination System
LCP	local control panel			
LEL	lower explosive limit		NPSHR	net positive suction head requirements
LFB	laboratory-fortified blank (standard)		NVS	nonvolatile solids
LFM	laboratory-fortified matrix (spike)		NVSS	nonvolatile suspended solids
LFMD	laboratory-fortified matrix duplicate (spike duplicate)		O&M	operation and maintenance
			O_3	ozone
LIMS	laboratory information management system		ODE	opposite drive end
LRA	locked rotor amperes		ODP	open drip-proof
m	meter		ORP	oxidation–reduction potential
m/s	meters per second		OSHA	Occupational Safety and Health Administration
m^3	cubic meters			
m^3/d	cubic meters per day		OTE	oxygen-transfer efficiency
m^3/kg	cubic meters per kilogram		OTR	oxygen-transfer rate
m^3/s	cubic meters per second		OUR	oxygen-uptake rate
mA	milliamperes or milliamps		PAM	polyacrylamide
MDL	method detection limit		PAO	phosphate accumulating organism
mg	milligram		Pa	Pascal
$Mg(OH)_2$	magnesium hydroxide		PEL	permissible exposure limit
mg/g	milligrams per gram		PF	power factor
mg/kg	milligrams per kilogram		PFRP	process to further reduce pathogens
mg/L	milligrams per liter		pH	inverse log of the hydrogen ion concentration
mgd	million gallons per day			
mil	1/1000 of an inch			
mil. gal	million gallons		PID	proportion, integral, derivative

PLC	programmable logic controller	s.u.	standard units (pH)
PM motor	permanent magnet motor	SVI	sludge volume index
PO_4	orthophosphate	TDS	total dissolved solids
POR	preferred operating range	TEBC	totally enclosed blower-cooled
ppb	parts per billion	TEFC	totally enclosed fan-cooled
PPE	personal protective equipment	TENV	totally enclosed non-ventilated
ppm	parts per million	TEXP	totally enclosed explosion proof
P	power	TKN	total Kjeldahl nitrogen
ppt	parts per trillion	TNVS	total nonvolatile solids
PQL	practical quantitation limit	TP	total phosphorus
PRV	pressure relief valve	TS	total solids
psi	pounds per square inch	TSS	total suspended solids
psig	pounds per square inch gauge	TVS	total volatile solids
PSRP	process to significantly reduce pathogens	TVSS	total volatile suspended solids
PTFE	polytetrafluoroethylene	TWAS	thickened waste activated sludge
PVC	polyvinyl chloride	U.S. EPA	United States Environmental Protection Agency
QA/QC	quality assurance and quality control		
RAS	return activated sludge	U.S.	United States
RBC	rotating biological contactor	UBOD	ultimate biochemical oxygen demand
RPD	relative percent difference	UEL	upper explosive limit
rpm	revolutions per minute	UPS	uninterruptible power supply
R	resistance	V	volts
RTU	remote terminal units	V	volume; may be shown as volume 1 (V_1), volume 2 (V_2), etc.
s	second		
SAE	standard aeration efficiency	VA	volatile acids
SAE	Society of Automotive Engineers	VA/ALK	volatile acids to alkalinity ratio
sBOD	soluble biochemical oxygen demand	VAR	vector attraction reduction
$sBOD_5$	5-day soluble biochemical oxygen demand	VAR	volts-amps reactive
SCADA	supervisory control and data acquisition	VFA	volatile fatty acids
scfm	standard cubic feet per minute	VFD	variable-frequency drive
SDS	safety data sheet, formally called material safety data sheet (MSDS)	VOC	volatile organic compound
		VS	volatile solids
SLR	solids loading rate	VSR	volatile solids reduction
sm^3/h	standard cubic meters per hour	VSS	volatile suspended solids
SO_2	sulfur dioxide	W	watts
SOR	surface overflow rate	W/m^3	watts per cubic meter
SOUR	specific oxygen uptake rate	WAS	waste activated sludge
SPAM	solution polyacrylamide	WRRF	water resource recovery facility
SPDES	State Pollutant Discharge Elimination System	Ω	ohms
SRT	solids retention time	μS	microseimens
SSV_5	settled sludge volume at 5 minutes		

Answer Keys

CHAPTER 1

SOLIDS CHARACTERISTICS

1. True
2. True
3. False. Secondary sludge tends to be more dilute than primary sludge.
4. B
5. D
6. C
7. A
8. B
9. B
10. A
11. 1 = c, 2 = e, 3 = a, 4 = b, 5 = d
12. B

SOLIDS HANDLING

1. False. Solids require additional treatment to inactivate and reduce pathogens and meet VAR.
2. True
3. False. Contaminants are generally not harmful to operators.
4. D. Only settleable particles can be captured in the sludge.
5. C
6. A
7. C
8. Reduce volume, reduce mass, reduce percentage of organic material, inactivate or reduce the number of pathogens, and meet regulatory requirements.

REDUCING VOLUME

1. False. Thick primary sludge may contain 8% solids, but will still be more than 90% water.
2. True
3. False. Vicinal water and water of hydration cannot be removed with mechanical methods.
4. C
5. D

International Standard units

$$kg/d = \frac{[(\text{Concentration, mg/L})(\text{Volume, m}^3/\text{d})]}{1000}$$

$$kg/d = \frac{[(280 \text{ mg/L})(18\ 925 \text{ m}^3/\text{d})]}{1000}$$

$$kg/d = 5299$$

U.S. customary units

$$lb/d = (\text{Concentration, mg/L})(\text{Volume, mgd})(8.34)$$

$$lb/d = (280 \text{ mg/L})(5 \text{ mgd})(8.34)$$

$$\frac{lb}{d} = 11\ 676$$

6. A

International Standard units

Step 1—Find the mass of TSS in the influent.

$$\text{kg/d} = \frac{[(\text{Concentration, mg/L})(\text{Volume, m}^3/\text{d})]}{1000}$$

$$\text{kg/d} = \frac{[(315 \text{ mg/L})(11\ 355 \text{ m}^3/\text{d})]}{1000}$$

$$\text{kg/d} = 3577$$

Step 2—Find the mass of TSS going into the sludge.

$$\text{Percent} = \frac{\text{Part}}{\text{Whole}} \times 100$$

$$40 = \frac{\text{Part}}{3577 \text{ kg/d}} \times 100$$

$$\text{Part} = 1431 \text{ kg/d}$$

U.S. customary units

Step 1—Find the mass of TSS in the influent.

$$\text{lb/d} = (\text{Concentration, mg/L})(\text{Volume, mgd})(8.34)$$

$$\text{lb/d} = (315 \text{ mg/L})(3 \text{ mgd})(8.34)$$

$$\frac{\text{lb}}{\text{d}} = 7881$$

Step 2—Find the mass of TSS going into the sludge.

$$\text{Percent} = \frac{\text{Part}}{\text{Whole}} \times 100$$

$$40 = \frac{\text{Part}}{7881 \text{ lb/d}} \times 100$$

$$\text{Whole} = 3152 \text{ lb/d}$$

7. C

$$\frac{5.3\%}{} \left| \frac{10\ 000 \text{ mg/L}}{1\%} \right| = 53\ 000 \text{ mg/L}$$

8. B

International Standard units

$$C_1 V_1 = C_2 V_2$$

$$\left(7000\frac{\text{mg}}{\text{L}}\right)(190 \text{ m}^3) = (3\%)(V_2)$$

$$\left(7000\frac{\text{mg}}{\text{L}}\right)(190 \text{ m}^3) = \left(30\ 000\frac{\text{mg}}{\text{L}}\right)(V_2)$$

$$V_2 = 44.3 \text{ m}^3$$

U.S. customary units

$$C_1V_1 = C_2V_2$$

$$\left(7000\frac{mg}{L}\right)(50\ 000\ gal) = (3\%)(V_2)$$

$$\left(7000\frac{mg}{L}\right)(50\ 000\ gal) = \left(30\ 000\frac{mg}{L}\right)(V_2)$$

$$V_2 = 11\ 667\ gal$$

Note: A separate solution is not needed for International Standard units in this case.

9. A

International Standard units

Step 1—Convert percent solids to mg/L. Units must match those given in the equation.

$$\frac{8\% | 10\ 000\ mg/L|}{1\%} = 80\ 000\ mg/L$$

Step 2—Use the mass equation to find volume.

$$kg/d = \frac{[(Concentration,\ mg/L)(Volume,\ m^3/d)]}{1000}$$

$$1854.65\ kg/d = \frac{[(80\ 000\ mg/L)(Volume,\ m^3/d)]}{1000}$$

$$Volume,\ m^3/d = 23.18$$

U.S. customary units

Step 1—Convert percent solids to mg/L. Units must match those given in the equation.

$$\frac{8\% | 10\ 000\ mg/L|}{1\%} = 80\ 000\ mg/L$$

Step 2—Use the mass equation to find volume.

$$lb/d = (Concentration,\ mg/L)(Volume,\ mgd)(8.34)$$

$$4086.6\ lb/d = (80\ 000\ mg/L)(Volume,\ mgd)(8.34)$$

$$Volume,\ mgd = 0.006125$$

Step 3—Convert mgd to gpd.

$$\frac{0.006125\ mil.\ gal | 1\ 000\ 000\ gal|}{d \qquad | 1\ mil.\ gal|} = 6125\ gpd$$

10. A
11. C
12. A
13. B
14. C

SLUDGE CONDITIONING

1. True
2. False. Inorganic conditioners add mass. For every 1 kg (lb) of ferric or lime added to condition sludge, the sludge mass increases by at least 1 kg (lb).
3. False. Viscosity is not an indicator of polymer content.
4. True
5. True
6. C
7. A
8. C
9. B
10. C
11. B
12. C. Chemical sludge tends to be positively charged.
13. B
14. C
15. A
16. B
17. D
18. A
19. C

LABORATORY TESTING

1. False. Operators should target the smallest dose required to achieve good results.
2. False. Polymer costs must be balanced against sludge transport costs.
3. False. A metric ton is equivalent to 1000 kg, or 2204 lb.
4. A
5. B
6. C
7. C
8. C. While the fourth jar test gave the best results, it required much more chemical than the third jar test, which also had good results.
9. B
10. A

REDUCTION REQUIREMENTS

1. False. Methane is produced in anaerobic digesters. Endogenous respiration occurs in aerobic digesters.
2. True
3. True
4. True
5. B
6. C
7. A
8. A
9. D
10. C

$$\text{Reduction of Volatile Solids, \%} = \left[\frac{(VS_{in} - VS_{out})}{VS_{in} - (VS_{in} \times VS_{out})} \right] \times 100$$

$$\text{Reduction of Volatile Solids, \%} = \left[\frac{(0.83 - 0.71)}{0.83 - (0.83 \times 0.71)} \right] \times 100$$

$$\text{Reduction of Volatile Solids, \%} = \left[\frac{(0.12)}{0.83 - (0.5893)} \right] \times 100$$

$$\text{Reduction of Volatile Solids, \%} = 49.8$$

CHAPTER 2

PURPOSE AND FUNCTION OF THICKENING
1. D
2. C
3. D

GRAVITY THICKENING
1. False. Gravity thickeners perform best when thickening primary sludge or lime conditioned sludge.
2. D
3. False. Typically, gravity thickeners have sloped bottoms that help move the thickened sludge to the center hopper.
4. A) Rake arm
    B) Scraper blade
    C) Picket
    D) Feedwell
5. C
6. A
7. True
8. True
9. False. Primary fed gravity thickeners have a detention time of 24 to 48 hours.

DISSOLVED AIR FLOTATION THICKENING
1. True
2. D
3. True
4. B
5. B
6. False. Floats are typically designed to be 200 to 610 mm (8 to 24 in.) deep.
7. C
8. False. Low dissolved air can be caused by the air compressor system being off or set too low or because the educator is clogged.
9. True
10. C

GRAVITY BELT THICKENERS
1. D
2. D
3. B
4. False. The filtrate should be clear in color.
5. B
6. True

CENTRIFUGE THICKENING
1. True. While it is not necessary to use polymer to condition the feed sludge to a centrifuge depending on the desired solids concentration, it is more common than not to do so.
2. C
3. True. The centrate color should be mostly clear. If the centrate becomes dark it is an indication of poor solids recovery in the unit and that should be addressed.
4. True
5. False. The bowl and scroll operate at slightly different speeds, which allows for the sludge to be thickened.
6. B
7. A

ROTARY DRUM THICKENERS

1. True
2. B
3. D
4. A
5. B
6. True
7. C
8. True. Effluent can be used as a water supply for the wash water system but strainers may have to be used to prevent clogging of the spray nozzles.

POLYMER SYSTEMS

1. C
2. A
3. C
4. C
5. A
6. B
7. 5, 6, 1, 2, 4, 3
8. C
9. D

SAFETY CONSIDERATIONS

1. False. Equipment should always be appropriately locked out/tagged out before any maintenance being performed.
2. True
3. C
4. False. A negative pressure is maintained to allow for air to be pulled into the cover but to prevent fouled air from leaking out.
5. True

CHAPTER 3

BIOLOGICAL TREATMENT FUNDAMENTALS REVIEW

1. D
2. A
3. C
4. D
5. A

THEORY OF OPERATION

1. True
2. False. Aerobic digesters have the same microorganisms as activated sludge and other aerobic treatment processes.
3. True
4. False. Nitrifying bacteria obtain their energy (food) from ammonia. Endogenous respiration ensures a steady supply of ammonia. They are able to live and grow in the digester.
5. True
6. False. Primary clarifiers are typically paired with anaerobic digesters. Exceptions include primary clarifiers upstream of fixed-film processes.
7. C
8. A
9. D
10. A
11. C
12. B
13. B
14. C

15. D
16. A
17. B
18. A
19. C
20. B
21. A
22. B

DESIGN PARAMETERS AND EXPECTED PERFORMANCE

1. True
2. False. Supernatant may also collect between layers of floating and settled sludge.
3. False. Fecal coliforms are typically not pathogenic. They are indicator organisms.
4. A
5. B
6. C
7. C
8. B
9. A. The solids in the supernatant also contribute to the amount of total nitrogen and phosphorus in the supernatant. Organic solids contain some percentage of both nitrogen and phosphorus.

EQUIPMENT

1. A
2. C
3. D
4. B
5. C

HYDRAULIC DETENTION TIME AND SOLIDS RETENTION TIME

1. True
2. B
3. B
4. D
5. C
6. B
7. C
8. D
9. D
10. C
11. A
12. B
13. A

OTHER PROCESS VARIABLES

1. True
2. False. Volatile solids are used to calculate the loading rate.
3. True
4. False. Volatile solids loading rate is based on fill volume, not digester capacity.
5. False. While related, one cannot be used to predict the other.
6. B. The digester dimensions are not needed to solve the problem.
7. A
8. D
9. B
10. A
11. C
12. B

13. B. Degree-days are always calculated using degrees Celsius.
14. B
15. C
16. D
17. A
18. B
19. B. In practice, many systems can't achieve higher than 2% TS.
20. D
21. A
22. B

PROCESS CONTROL
1. False. Most digesters decant periodically.
2. A
3. B
4. D
5. B. The three-normal equation can be used for all types of blending problems.
6. C
7. A

OPERATION
1. False. Start the digester with just enough feed sludge to cover the diffusers. Add influent or effluent to raise the water level up when using surface aerators.
2. True
3. C
4. A
5. B
6. C
7. A

MONITORING, MAINTENANCE, TROUBLESHOOTING, AND SAFETY
1. True
2. True
3. True
4. False. Always practice good personal hygiene to prevent illness.
5. B
6. A
7. C
8. D
9. A
10. C
11. B
12. B

CHAPTER 4

THEORY OF OPERATION
1. True
2. True
3. False. Anaerobic digesters are typically fed with a mixture of primary and secondary sludge.
4. False. Performance depends on feed sludge biodegradability.
5. False. Methanogens can also combine carbon dioxide and hydrogen gas to produce methane.
6. C
7. A
8. B

9. D
10. A
11. C
12. A
13. B
14. C
15. A
16. C

DESIGN PARAMETERS AND EXPECTED PERFORMANCE

1. B
2. D
3. B
4. A
5. C
6. B
7. C
8. D
9. A
10. B

EQUIPMENT

1. A
2. B
3. False. The exterior membrane is always fully inflated by fans that maintain constant pressure by blowing ambient air between the two membranes.
4. False. An ideal digester maintains complete mix flow throughout the tank.
5. B
6. C
7. False. Microorganisms in a digester are sensitive to temperature changes. Heating systems are designed to maintain digester temperature as constant as possible, with a typical target of no more than 0.6 °C (1 °F) of temperature change within a 24-hour period to help minimize the risk of foaming events.
8. C
9. B
10. B
11. C
12. B
13. D
14. True
15. C
16. A
17. B
18. A
19. B
20. D

PROCESS VARIABLES AND PROCESS CONTROL

1. True
2. False. As alkalinity increases, pH change is resisted.
3. False. VFA concentrations up to 8000 mg/L are tolerated by methanogens.
4. True
5. C
6. B
7. A
8. C
9. C

10. C
11. B
12. B
13. D
14. C
15. A
16. C
17. B
18. D
19. A
20. B
21. D
22. C
23. A

OPERATION

1. False. The readiness of all mechanical equipment should be checked before process fluid is placed in the tank. Installation checks, electrical connections inspections, and piping connection checks should be made and motors should be bumped.
2. B
3. True
4. False. Digesters perform better when they are fed a constant amount of VS continuously.
5. C
6. A
7. B
8. False. The heating and mixing systems must be turned off for decanting. Because decanting is usually performed in secondary digesters, there is less biological activity and the temperature change has little effect.
9. D
10. C
11. C
12. True

DATA COLLECTION, SAMPLING, AND ANALYSIS AND MAINTENANCE

1. A
2. A
3. D
4. B
5. C
6. B
7. C

SAFETY CONSIDERATIONS

1. True
2. B
3. False. Normal concentrations of oxygen in dry air are just under 21%. Areas with less than 19.5% oxygen are considered deficient.
4. A
5. C
6. D

CHAPTER 5

PURPOSE AND FUNCTION OF DEWATERING

1. False. Dewatering reduces the amount of water in the sludge, but has no effect on the amount of solids or total mass.
2. False. Chemical sludges tend to stick to the belts and clog the pores of the belt material.
3. True
4. True
5. False. Chemical conditioning is most often required in mechanical sludge dewatering.

CENTRIFUGES FOR DEWATERING

1. True
2. False. Dewatering takes place throughout the entire length of the centrifuge.
3. A
4. A
5. D
6. B
7. False. Centrifuges are effective at dewatering all types of solids.
8. True
9. False. Increasing the feed rate to a dewatering centrifuge will create a wetter cake with lower percent solids.

BELT FILTER PRESSES

1. True
2. False. BFPs work well to dewater all types of sludges with the exception of chemical sludges because they are stickier.
3. False. Dilute feed solids require more time on the gravity drainage section of the belt and some BFPs are designed with longer gravity sections to accommodate for this.
4. D
5. True
6. True
7. C
8. B
9. C
10. True
11. A
12. C
13. C
14. B
15. False. Dilute chlorine solution can be used to clean the belts of a belt filter press.
16. True
17. False. Belt filter presses are useful in dewatering all solids with the exception of chemical and industrial solids.

ROTARY PRESSES

1. A
2. False. A conditioning chemical such as polymer is used to flocculate the solids and allow water to be released during the filtration portion of the rotary press.
3. C
4. B
5. True
6. C
7. C
8. D
9. B
10. A
11. C
12. D
13. B

SCREW PRESSES

1. B
2. False. Slow rotational speed is an advantage of screw press operations because it uses less power than other dewatering devices.
3. False. Dewatered sludge concentrations from a screw press typically range between 12% and 40% TS.
4. True
5. C
6. True
7. False. Screw presses rotate at a maximum of 2 rpm, while centrifuges operate as high as 3000 rpm.
8. A
9. True

10. C
11. False. Screw presses operate at slow rotational speeds.
12. True

DRYING BEDS

1. B
2. D
3. C
4. False. Drying beds are most often used to dewater digested solids.
5. True
6. D
7. False. Some sand is removed every time sludge is removed from a sand drying bed and it will need to be replaced over time.
8. C
9. False. Drying beds should be cleaned of debris between each sludge application.

OTHER DEWATERING PROCESSES

1. C
2. D
3. A

SLUDGE CAKE CONVEYORS

1. D
2. True
3. B
4. C

DEWATERED SLUDGE STORAGE AND HAULING

1. D
2. False. Dewatered biosolids do not flow well and are most often stored in buildings or short-term locations like semi-trailers.

ODOR CONTROL AND SAFETY

1. True
2. False. Safety interlocks should be supplied with any dewatering system installation.
3. True

CHAPTER 6

WHAT IS ELECTRICITY?

1. B
2. A
3. C
4. True
5. D
6. B
7. D
8. B
9. True
10. False. Three things are required to generate electricity: a magnetic field, a conductor, and relative movement. Electricity will only be generated if either the conductor or the magnet is moving.
11. True
12. C

PROPERTIES OF ELECTRICITY

1. A
2. C

3. False
4. C
5. A
6. C
7. B
8. A
9. D
10. False
11. A = 2, B = 5, C = 4, D = 3, E = 1

RELATIONSHIPS BETWEEN PROPERTIES OF ELECTRICITY
1. A
2. B
3. A
4. A

PARALLEL AND SERIES CIRCUITS
1. False. The current is the same through a series circuit, not a parallel circuit.
2. C
3. True
4. A
5. True

DIRECT CURRENT AND ALTERNATING CURRENT
1. B
2. D
3. A
4. False. Electrons pass energy from one to the other. Energy moves, not electrons.
5. C
6. D
7. False. Electricity doesn't typically travel through the equipment grounding conductor (wire).
8. True
9. B
10. True
11. B
12. C

TRANSFORMERS AND SHORT CIRCUITS AND GROUND FAULTS
1. False. Direct current does not cycle.
2. C
3. True
4. D
5. False. An overload does not cause a surge in current.
6. B

MOTORS
1. D
2. A
3. B
4. True
5. B
6. C
7. True
8. B
9. D
10. False. The operating speed must be less than the synchronous speed.

NAMEPLATE INFORMATION

1. B
2. C
3. False. Ampere draw is highest at startup.
4. B
5. A
6. B
7. True
8. B
9. D

MOTOR CONTROL AND ROUTINE MOTOR OPERATION AND MAINTENANCE

1. A
2. D
3. B
4. C
5. A
6. C
7. A
8. True
9. False. All three leads must be checked and compared against each other.
10. B
11. C
12. A
13. C
14. D

ELECTRICAL PRINTS AND DISCONNECTS AND MOTORS

1. A
2. C
3. D
4. B
5. D

PUSH BUTTONS AND SWITCHES

1. D
2. C
3. B
4. A
5. D
6. C

RELAY COILS AND CONTACTS

1. A
2. D
3. C
4. D
5. B

SAFETY-RELATED WORK PRACTICES

1. A
2. C
3. A
4. B

CHAPTER 7

CLASSIFICATION OF PUMPS

1. B
2. C
3. True

CENTRIFUGAL PUMPS

1. C
2. C
3. C
4. B
5. C
6. C
7. C
8. A
9. D
10. B
11. D
12. C
13. A
14. C
15. B
16. B
17. C
18. D
19. C
20. D
21. C
22. A
23. B
24. B
25. D
26. C
27. B
28. B
29. D
30. C

POSITIVE DISPLACEMENT PUMPS

1. C
2. False. Unlike centrifugal pumps, positive displacement pumps are capable of generating nearly full design pressure over the full range of flow.
3. B
4. C

DIAPHRAGM AND DISC PUMPS

1. C
2. B
3. D
4. D
5. C
6. C
7. A
8. D

PLUNGER PUMPS

1. C
2. B
3. D
4. B
5. C
6. B
7. D
8. B

PISTON PUMPS

1. D
2. D
3. B
4. C
5. D
6. A
7. C
8. D

ROTARY LOBE PUMPS

1. C
2. C
3. D

PROGRESSING CAVITY PUMPS

1. C
2. False. Progressing cavity pumps can pump equally effectively in either direction by wiring the motor to rotate in the opposite direction.
3. B
4. D

SCREW PUMPS

1. C
2. B

PERISTALTIC PUMPS

1. B
2. D

LIFT STATIONS/PUMPING STATIONS

1. B
2. False. As more flow is added from other lift stations and sewers, lift stations typically increase in size downstream.
3. B
4. C
5. D
6. B
7. D
8. A. Even when stairs and continuous ventilation systems are provided, care should be taken to check for available oxygen and explosive gas.
9. B
10. D
11. B
12. C
13. A
14. D

15. B
16. B
17. C
18. A
19. C
20. D
21. C
22. D
23. D
24. A
25. C
26. C
27. B
28. D
29. D
30. A
31. C
32. B

CHAPTER 8

AERATION THEORY

1. D
2. B
3. C
4. B
5. False. The oxygen within the bubble is not dissolved. When the bubble bursts, the air and the gaseous oxygen within it are released back into the atmosphere. Dissolved oxygen exists as O_2 molecules in the water.
6. True. A specific volume of air as small bubbles has a significantly higher surface area than the same volume of air as big bubbles. The additional surface area provides more opportunity for the gaseous oxygen to dissolve into the liquid.
7. D

TYPES OF AERATION SYSTEMS

1. False. Because mechanical aerators are typically less complex than diffused aeration systems, they are typically less expensive to purchase.
2. True. The various equipment components, monitoring instrumentation, and controls often included in diffused air systems provide degrees of adjustment and more precise control than simpler mechanical aerator systems.

DIFFUSED AIR SYSTEMS

1. The components of diffused aeration systems include blowers, diffusers, air valves, controls, DO analyzers, and air downcomers (droplegs).

BLOWERS

1. A
2. D
3. C
4. D
5. B
6. True. The wastewater market is trending toward "plug and play" skid-mounted blower packages that come ready to be powered up and put into operation; controls and accessories are included with the blower and motor.

BLOWER SELECTION CRITERIA

1. C
2. C
3. A

4. True. Blowers produce more noise than most other types of equipment at a WRRF. Because hearing protection is typically not needed throughout a WRRF, it needs to be specifically assessed if people should use it when working in the vicinity of the blowers.
5. Important blower selection criteria: maintenance requirements, life-cycle costs, footprint, noise levels, energy efficiency, controls options, and performance requirements.
6. False. The length of noise exposure time has a significant effect on hearing damage.
7. A

COMMON BLOWER ACCESSORIES

1. A
2. Paper filter elements are more efficient than wire filter element, cheaper to purchase than cloth filter elements, and less durable than cloth filter elements.
3. C
4. Rubber couplings are used to connect blower inlets and outlets to piping to prevent blower vibration from traveling through the air piping.
5. C
6. True. Many positive displacement blowers are arranged in this manner and driven by V-belts.
7. D
8. A
9. B
10. Efficiencies and speeds
11. B
12. C

CENTRIFUGAL BLOWERS

1. D
2. C
3. C
4. A
5. False
6. B
7. The inlet valve is throttled too much, the discharge isolation valve is closed by mistake, and the diffusers are severely fouled.
8. C
9. B
10. B

INTEGRALLY GEARED SINGLE-STAGE CENTRIFUGAL BLOWERS

1. C
2. True. Integrally geared single-stage centrifugal blowers can achieve isentropic efficiencies up to 80%.

MULTISTAGE CENTRIFUGAL BLOWERS

1. D
2. A

HIGH-SPEED-DRIVEN (TURBO) BLOWERS

1. C
2. True. Air foil and magnetic bearings create an air gap between the rotor and the stator.
3. B
4. The benefits of plug-and-play systems include single source responsibility, factory testing of the system, simplified installation, and simplified startup.
5. False. Turbo blowers have a higher peak efficiency than most other blowers used at WRRFs.
6. C
7. D

POSITIVE DISPLACEMENT BLOWERS

1. False. Positive displacement blowers are known as *constant volume* machines.
2. D

3. A
4. True. This characteristic makes positive displacement blowers ideal for operating in variable level tanks.

HYBRID ROTARY SCREW BLOWERS
1. A
2. False. Rotary screw blowers are often belt-driven or direct-drive.

AIR DISTRIBUTION PIPING AND VALVES
1. A
2. C
3. True. This improves control. If movement is allowed near 100% open, lower velocities do not provide significant change; movement near 100% closed causes excessive change.
4. Air valves are used to create pressure differential, isolate an air grid, and control the quantity of air going to a reactor or individual air grid.

DIFFUSERS
1. B
2. B
3. B
4. C
5. D
6. C
7. B

AERATION SYSTEM CONTROL
1. C
2. B
3. A

JET AERATION SYSTEMS
1. True. Very small bubbles are created by the interaction of the air and water within the nozzles at the point where the process flow reenters the tank.
2. A

SURFACE AERATORS
1. A
2. Typical preventive maintenance tasks for surface aerators include the following: change the oil in the drive and grease the motor.

HORIZONTAL ROTORS
1. C
2. C

MIXER AERATORS
1. D
2. Mixer aerator components include submerged mixer body, sparger ring, separate air supply, and platform-mounted motor.
3. A
4. B
5. False. Mixer aerators do not use diffusers, so no fouling occurs.
6. False. Mixer aerators do not have inlet guide vanes. The speed of the mixer and the airflow rate to the sparger ring can be adjusted.

CHAPTER 9

FLOW MEASUREMENT
1. True
2. False. Differential head flow meters can only be used with full pipes.

3. A
4. B
5. D
6. B

SAMPLING

1. True
2. False. To know if a sample is representative, one must also have an approximation of the true average.
3. True
4. True
5. C
6. B
7. B
8. C
9. A
10. A

PURPOSES OF SAMPLING

1. True
2. False. All laboratory methods have limits on their sensitivity. If the detection limit is 2 mg/L, a concentration of 1 mg/L would be reported as <2 mg/L.
3. False. The discharge permit only includes reporting requirements for the influent and effluent. It does not include requirements for process control monitoring.
4. True
5. B
6. D

SAMPLE AND MEASUREMENT COLLECTION

1. True
2. True
3. False. Samples should be well mixed and poured quickly to prevent the settling of solids during transfer.
4. False. Samples may be collected over a period of time in two different ways. The operator could collect the same volume of sample at fixed time intervals (e.g., every 2 hours) while also taking a flow measurement reading. At the end of sampling, the average flow during the sampling event will be known and samples can be aliquoted appropriately. Alternatively, the operator can collect a fixed volume of sample per unit of flow (e.g., 300 mL per ML [mgd]).
5. B
6. A
7. A
8. B
9. D
10. C
11. D
12. A

SAMPLING EQUIPMENT

1. True
2. False. Although some sample types require special cleaning, most of the parameters in wastewater like TSS, BOD, CBOD, and TKN do not require acid washing.
3. True
4. True
5. B
6. D
7. C
8. C
9. B
10. B

11. A
12. D

SAMPLE HANDLING

1. True
2. False. Composite samples may not be collected for parameters where the hold time is less than the total time of the composite. In this case, chlorine residual has a 15-minute hold time. U.S. EPA defines a composite sample as any sample collected over longer than 15 minutes. By definition, residual chlorine samples must be grab samples.
3. True
4. True
5. C
6. C
7. D
8. B
9. A

QUALITY ASSURANCE AND QUALITY CONTROL

1. True
2. False. Blanks should be randomized to determine if all glassware is being cleaned adequately.
3. False. While replicates are averaged together, laboratory and field duplicates are never averaged together with the original sample result.
4. A = 2, B = 1, C = 5, D = 4, E = 3
5. A
6. B
7. B
8. B
9. D
10. C
11. A
12. B
13. C
14. D
15. C
16. B
17. A
18. B
19. D
20. B
21. C
22. B
23. B
24. C

ANALYTICAL METHODS

1. False. Colorimeters and spectrophotometers with internally stored calibration curves only require a single standard.
2. True
3. C
4. A
5. B
6. D
7. B
8. C
9. A
10. D

pH (HYDROGEN ION CONCENTRATION)

1. False. Samples with pH > 7 are basic.
2. True
3. A
4. B
5. C
6. D

TOTAL ALKALINITY OF WASTEWATER AND SLUDGE

1. True
2. True
3. D
4. C
5. A
6. B
7. A
8. B

TOTAL SUSPENDED SOLIDS (NONFILTERABLE RESIDUE)

1. True
2. True
3. False. Dry areas provide a path for air movement. The filter disk will not dry sufficiently and may stick to the aluminum pan.
4. C
5. D
6. B
7. A
8. C
9. B
10. B
11. C
12. C
13. B
14. C
15. A
16. D

BIOCHEMICAL OXYGEN DEMAND

1. False. The BOD test measures the amount of oxygen consumed, which can then be used to estimate the organic strength.
2. True
3. True
4. C
5. D
6. B
7. A
8. C
9. D
10. C
11. A
12. C
13. B
14. C
15. D
16. A
17. B

18. C
19. D
20. B

CHAPTER 10

PURPOSE AND FUNCTION
1. B
2. True
3. False. If the chemical is used to control environmental conditions, it is supporting another treatment process.

THEORY OF OPERATION
1. C
2. False. Chemicals can degrade over time and lose effectiveness or become difficult to handle. Storage should be sized to rotate chemical stock every few weeks.
3. D
4. True

DESIGN PARAMETERS
1. C. 1.5 times the delivery size is 25 500 L (6750 gal), but, for 2 weeks (14 days) of operation at 2650 L/d (700 gpd), 37 000 L (9800 gal) is required.
2. D

EQUIPMENT
1. B
2. B
3. C
4. False. Regardless of volume, spill containment is a required part of chemical storage.
5. D
6. False. Chemical pumps are a potential source of leaks and should be located within the containment area, but elevated so they are not damaged in the event of a chemical spill.
7. False. Returning undiluted or un-neutralized chemical rapidly and directly to the facility influent is likely to cause major process upsets.
8. False. Recovery of the chemical is preferred for financial reasons.
9. D
10. D
11. True
12. A
13. B

RECEIVING FACILITIES
1. True
2. A
3. B

CHEMICAL TRANSFER
1. False. A chemical day tank may be filled several times a day to once every few days.
2. D
3. B
4. C
5. False. Centrifugal pumps are capable of higher flows than diaphragm pumps.
6. D
7. B

DRY CHEMICAL SYSTEMS

1. A
2. True
3. C
4. D
5. False. Gravimetric feeders are more precise because they account for differences in density.
6. B
7. D
8. False. Soluble chemicals form solutions, whereas insoluble chemicals form slurries or suspensions.
9. True
10. A
11. B
12. B

LIQUID CHEMICAL SYSTEMS

1. B
2. False. The vent pipe should not have a valve and should always remain open to avoid pressurizing the tank during the fill cycle and to relieve vacuum as the tank is drawn down.
3. False. Intermediate bulk containers are larger than drums.
4. B
5. C
6. C
7. C
8. A
9. True
10. A

GAS CHEMICAL SYSTEMS

1. B
2. False. Cylinders have only a top valve for withdrawing gas.
3. A
4. False. In a vacuum system, air will be drawn into the system through any leaks. In a pressure system, gas will be forced out of any leaks into the work area.

CHEMICAL APPLICATION

1. B
2. A
3. B
4. a = 2, b = 1, c = 3, d = 4

PROCESS VARIABLES

1. False. The technical data sheet only includes information on the chemical composition and properties, not safety information.
2. False. The specific gravity of water is 1. A chemical with a specific gravity of less than 1 is less dense than water.
3. C
4. C

CALCULATING THE MASS OF ACTIVE CHEMICAL IN NEAT CHEMICAL AND ELEMENT OR ION OF USE IN THE ACTIVE CHEMICAL

1. D
2. D
3. C
4. C

CHEMICAL DOSE

1. D
2. False

3. B
4. A
5. B
6. C

USING ABC EQUATIONS FOR CHEMICAL FEED PUMP SETTINGS

1. D
2. False. The specific gravity will be 1.1.
3. D
4. D
5. A

REACTION TIME

1. False. Zero-order reactions do not depend on the concentration of reactants and proceed at a constant rate.
2. False. Rate constants are directly proportional to the reaction rate. Higher rate constants mean higher rates.
3. B
4. True

SIDE REACTIONS

1. C
2. False. Competition from other compounds and non-ideal environmental and reactor conditions cause the actual dose needed to be higher than stoichiometric doses.

PROCESS CONTROL

1. A
2. D
3. False. Flow-paced control, as referenced in the question, is one option for chemical feed control. Constant speed or ratio modes should also be considered.
4. C
5. False. Constant speed control will not respond to variations in influent flow or load and will produce a variable quality effluent.
6. D
7. A

CHEMICAL SAFETY

1. D
2. B
3. D
4.

Information	SDS Section
a. LD_{50} values	Toxicological Information
b. List of known dangerous reactions	Stability and Reactivity
c. Emergency number	Identification
d. Chemical formula	Composition/Information on Ingredients
e. Recommended PPE	Exposure Controls/Personal Protection
f. Exposure routes	First Aid Measures
g. Pictogram	Hazard Identification
h. Storage conditions	Handling and Storage
i. Materials and methods of cleanup	Accidental Release Measures
j. Melting/freezing point	Physical and Chemical Properties
k. Suitable extinguishing media	Fire-Fighting Measures

5. A
6. C
7. False. Appropriate PPE must be worn whenever working around chemicals.

8. C

9. D

10. False. Overprotection increases the stress on the worker. Select an appropriate level of protection for the assessed risk.

11.

White	Special Hazard
Blue	Health
Red	Flammability
Yellow	Instability

12. False

13. A

Periodic Table

PERIODIC TABLE
group

1	2	3	4	5	6	7	8	9	10	11	12	13	14	15	16	17	18

Legend / key:

- element name ———▸ **HYDROGEN**
- atomic number ———▸ **1**
- chemical Symbol ———▸ **H**
- atomic weight (u) ———▸ 1,008

Alkali metal
Alkaline earth metal
Lanthanide
Actinide

Transition metal
Post-transition metal
Metalloid

Nonmetal
Halogen
Noble gas

● Solid
● Liquid
● Gas
● Unknown

Period 1

IA												IIIA	IVA	VA	VIA	VIIA	VIIIA
HYDROGEN 1 **H** 1,008																	HELIUM 2 **He** 4,003

Period 2

| LITHIUM 3 **Li** 6,941 | BERYLLIUM 4 **Be** 9,012 | | | | | | | | | | | BORON 5 **B** 10,81 | CARBON 6 **C** 12,01 | NITROGEN 7 **N** 14,01 | OXYGEN 8 **O** 16,00 | FLUORINE 9 **F** 19,00 | NEON 10 **Ne** 20,18 |

IIA

Period 3

| SODIUM 11 **Na** 22,99 | MAGNESIUM 12 **Mg** 24,31 | | | | | | | | | | | ALUMINIUM 13 **Al** 26,98 | SILICON 14 **Si** 28,09 | PHOSPHORUS 15 **P** 30,97 | SULFUR 16 **S** 32,07 | CHLORINE 17 **Cl** 35,45 | ARGON 18 **Ar** 39,95 |

IIIB IVB VB VIB VIIB VIIIB VIIIB VIIIB IB IIB

Period 4

| POTASSIUM 19 **K** 39,10 | CALCIUM 20 **Ca** 40,08 | SCANDIUM 21 **Sc** 44,96 | TITANIUM 22 **Ti** 47,87 | VANADIUM 23 **V** 50,94 | CHROMIUM 24 **Cr** 52,00 | MANGANESE 25 **Mn** 54,94 | IRON 26 **Fe** 55,85 | COBALT 27 **Co** 58,93 | NICKEL 28 **Ni** 58,69 | COPPER 29 **Cu** 63,55 | ZINC 30 **Zn** 65,39 | GALLIUM 31 **Ga** 69,72 | GERMANIUM 32 **Ge** 72,59 | ARSENIC 33 **As** 74,92 | SELENIUM 34 **Se** 78,96 | BROMINE 35 **Br** 79,90 | KRYPTON 36 **Kr** 83,80 |

Period 5

| RUBIDIUM 37 **Rb** 85,47 | STRONTIUM 38 **Sr** 87,62 | YTTRIUM 39 **Y** 88,91 | ZIRCONIUM 40 **Zr** 91,22 | NIOBIUM 41 **Nb** 92,91 | MOLYBDENUM 42 **Mo** 95,94 | TECHNETIUM 43 **Tc** (98,91) | RUTHENIUM 44 **Ru** 101,1 | RHODIUM 45 **Rh** 102,9 | PALLADIUM 46 **Pd** 106,4 | SILVER 47 **Ag** 107,9 | CADMIUM 48 **Cd** 112,4 | INDIUM 49 **In** 114,8 | TIN 50 **Sn** 118,7 | ANTIMONY 51 **Sb** 121,8 | TELLURIUM 52 **Te** 127,6 | IODINE 53 **I** 126,9 | XENON 54 **Xe** 131,3 |

Period 6

| CAESIUM 55 **Cs** 132,9 | BARIUM 56 **Ba** 137,3 | LANTHANUM 57 **La** 138,9 | HAFNIUM 72 **Hf** 178,5 | TANTALUM 73 **Ta** 180,9 | TUNGSTEN 74 **W** 183,9 | RHENIUM 75 **Re** 186,2 | OSMIUM 76 **Os** 190,2 | IRIDIUM 77 **Ir** 192,2 | PLATINUM 78 **Pt** 195,1 | GOLD 79 **Au** 197,0 | MERCURY 80 **Hg** 200,6 | THALLIUM 81 **Tl** 204,4 | LEAD 82 **Pb** 207,2 | BISMUTH 83 **Bi** 209,0 | POLONIUM 84 **Po** (210,0) | ASTATINE 85 **At** (210,0) | RADON 86 **Rn** (222,0) |

Period 7

| FRANCIUM 87 **Fr** (223,0) | RADIUM 88 **Ra** (226,0) | ACTINIUM 89 **Ac** (227,0) | RUTHERFORDIUM 104 **Rf** ---- | DUBNIUM 105 **Db** ---- | SEABORGIUM 106 **Sg** ---- | BOHRIUM 107 **Bh** ---- | HASSIUM 108 **Hs** ---- | | | COPERNICIUM 112 **Cn** ---- | | | | | | | |

Lanthanides

| CERIUM 58 **Ce** 140,1 | PRASEODYMIUM 59 **Pr** 140,9 | NEODYMIUM 60 **Nd** 144,2 | PROMETHIUM 61 **Pm** (144,9) | SAMARIUM 62 **Sm** 150,4 | EUROPIUM 63 **Eu** 152,0 | GADOLINIUM 64 **Gd** 157,3 | TERBIUM 65 **Tb** 158,9 | DYSPROSIUM 66 **Dy** 162,5 | HOLMIUM 67 **Ho** 164,9 | ERBIUM 68 **Er** 167,3 | THULIUM 69 **Tm** 168,9 | YTTERBIUM 70 **Yb** 173,0 | LUTETIUM 71 **Lu** 175,0 |

Actinides

| THORIUM 90 **Th** (232,0) | PROTACTINIUM 91 **Pa** (231,0) | URANIUM 92 **U** (238,0) | NEPTUNIUM 93 **Np** (237,0) | PLUTONIUM 94 **Pu** (239,1) | AMERICIUM 95 **Am** (243,1) | CURIUM 96 **Cm** (247,1) | BERKELIUM 97 **Bk** (247,1) | CALIFORNIUM 98 **Cf** (252,1) | EINSTEINIUM 99 **Es** (252,1) | FERMIUM 100 **Fm** (257,1) | MENDELEVIUM 101 **Md** (256,1) | NOBELIUM 102 **No** (259,1) | LAWRENCIUM 103 **Lr** (260,1) |